Charles Thomas Jacobi

The Printers' Handbook

Of Trade Recipes, Hints & Suggestions Relating to Letterpress and Lithographic

Printing, Bookbinding Stationery, Engraving, etc.

Charles Thomas Jacobi

The Printers' Handbook
Of Trade Recipes, Hints & Suggestions Relating to Letterpress and Lithographic Printing, Bookbinding Stationery, Engraving, etc.

ISBN/EAN: 9783337254902

Printed in Europe, USA, Canada, Australia, Japan

Cover: Foto ©berggeist007 / pixelio.de

More available books at **www.hansebooks.com**

The Printers' Handbook

OF

TRADE RECIPES, HINTS, & SUGGESTIONS

RELATING TO

LETTERPRESS AND LITHOGRAPHIC PRINTING

BOOKBINDING, STATIONERY

ENGRAVING, ETC.

WITH MANY USEFUL TABLES AND AN INDEX

COMPILED BY

CHARLES THOMAS JACOBI

Manager of the Chiswick Press ; Examiner in Typography to the City and Guilds of London Institute; Author of " Printing, a Practical Treatise on Typography ; " " The Printers' Vocabulary ; " "On the Making and Issuing of Books ; " etc.

Second Edition, Enlarged and Classified

LONDON

THE CHISWICK PRESS, 20 & 21, TOOKS COURT

CHANCERY LANE

1891

PREFACE TO THE ENLARGED AND CLASSIFIED EDITION.

THE compiler issues this Second Edition with a hope that it may prove as acceptable as at its first appearance. The facts that the first edition was soon exhausted, and continued inquiries were being made for the book, are the best excuse for its reappearance.

Much useful matter has been added, and an attempt has been made to classify it under the heads of departments or trades.

To have given more explicit acknowledgment would, perhaps, have been better, but, as intimated in the earlier preface, the same item is so frequently repeated in the different journals, and sometimes in a varying form, that it would have been difficult to have traced its origin. Owing to this repetition and variety in the mode of treatment, the reader's indulgence is requested if the compiler has failed in a few instances to discriminate between these duplicate fugitive pieces.

In addition to the journals named in the original preface, he is indebted to the following :—

BRITISH PRINTER.
BRITISH BOOKMAKER.
EFFECTIVE ADVERTISER.
PRINTING WORLD.

1891.

INTRODUCTORY TO THE FIRST EDITION.

IN putting this book before the Trade, the compiler offers his apologies and tenders his thanks to the proprietors of the various technical journals, both English and foreign, from which he has culled the bulk of these recipes, etc.; others are the result of his own practical experience. He trusts that this compilation—the result of many years' collection—will be found useful alike to the master and to the workman, for he believes no such work is at present in existence. It is hoped it will be of service to all interested or engaged in the art of printing and its allied trades. The want of such a book has long been felt, for, owing to the fact that the information is spread over a vast number of sources and repeated over and over again in different journals, there has always been a difficulty in turning up any particular subject when required. Further, the compiler has not had the opportunity of verifying all the matter contained in this work, so in many instances it has been reproduced exactly as it appeared from time to time.

If the demand should warrant a new edition at some future date, the compiler may be able to extend its usefulness, and he will be obliged by any hints, suggestions, or additional information suitable for the work being sent to him personally at the Chiswick Press, Tooks Court, Chancery Lane, London.

It is hoped the Index will be sufficient to find any particular subject, for it was found inconvenient to classify the multitude of items treated—at least for this initial edition.

Amongst others the compiler is indebted to the following English journals :—

PAPER AND PRINTING TRADES JOURNAL.

PRINTING TIMES AND LITHOGRAPHER.

PRINTING TRADES' DIARY AND DESK BOOK.

PRESS NEWS.

PRINTERS' REGISTER.

BRITISH AND COLONIAL PRINTER AND STATIONER.

1887.

CONTENTS.

THE PRINTERS' HANDBOOK

OF

RECIPES, HINTS, AND SUGGESTIONS.

COMPOSING ROOM.

ORIGIN of the Various Sizes of Type.—There were seven sizes of type. The first was called "prima," whence the name prime, but this sort is now termed two-line English. The second was "secunda," which is our double pica ; in France, great paragon. The third was "tertia," at present our great primer. Then there was the middle size, still called in German "mittel," but this is now our English. After these came the three sizes on the opposite side of the scale—pica, long primer, and brevier. In Germany the names secunda, tertia, and mittel are still retained. Pica in France and Germany is called "Cicero," because the works of that author were originally printed in it. English printers so styled it from being the type in which the ordinal or service-book of the Roman Church was originally set. This ordinal was first called the pica, or familiarly, pie. Bourgeois was so named because it was introduced into this country from France, where it was originally dedicated to the bourgeois or citizen printers of that capital. Brevier obtained its name from its having been first used in printing the breviary or Roman Catholic abbreviated church service-book. Nonpareil was so named because on its introduction it had no equal, being the smallest and finest type produced until that time. Pearl is of English origin. The French have a type of the same size which they call "Parisien." It was a smaller type than nonpareil, and was thought the pearl of all type. Diamond is another fancy name given to what was regarded at the time of its origin as

B

the greatest of letter-foundry achievements. One or two sizes besides have been made, and capriciously named by their respective producers. There is no doubt, however, that the best, because the most scientific and accurate, system of designing types is the French system—according to " points." At home here in England, our type-founders naturally object and hesitate to adopt an innovation of so revolutionary a character, hence the present time-honoured names are likely to hold their own in the terminology of the printing office.

Style of the House.—In large establishments it is best to print a few rules as to the method of doing certain things, as nothing causes so much friction between the reading and composing departments as a want of understanding in these matters. Peculiarities in pointing, spellings, and style are not always agreeable in different offices, and a variety of styles is perplexing to all concerned, and entails much labour both on readers and compositors.

Taste in Typography.—Though the word "taste" scarcely admits of the idea of variety of style, whether good or bad, it is pretty generally so understood. The proper use of it is to denote the sensation produced on the tongue as to distinguishing agreeable flavours. Figuratively, we may use the term in perceiving agreeable sounds, colour, or forms. In this sense a printer ought to try to educate himself to a high degree of excellence in discerning not only what is agreeable to his own eye, but to that of others. In other words, he should be able to appreciate a variety of tastes. The artist, whether he be a painter, photographer, or a printer, who can see beauty only in one style of art, or the musician who is always humming the same tune, can satisfy but few except himself. There may be but one perfect style of beauty in each department, and it may be possible for a man to attain to it, but it is certain that many more whose tastes may have had some cultivation are still unable to appreciate it. A printer, therefore, while he should strive to elevate his art by educating his customers up to his standard, should not seek to go so far above that they cannot follow.

Hints on Setting Pamphlet Covers.—The title for a pamphlet cover, without border, should be of plain face. Old style lower-case of Roman or italic will be most satisfactory for a short title of one or two lines. For a full-page cover-title select plain type. If a rule border is desired, select a rule which can be readily fitted with corners. Never cut a rule, nor make up a border, for a cover-title until you know what will be the exact size of the cover. When you know the size, arrange the border so that it will be equidistant on all sides from the edge and back of the cover. Always keep the border of a cover at good distance from the types of the title. Prefer borders of large pieces. Never make up a combination of small pieces without order. If a cover forme of four pages contains cuts or electrotyped advertisements of unequal size, have the four pages made up on galleys in pages of uniform size before they are laid on stone. Before making a margin, get a trimmed sheet of the cover paper. Find out from the foreman the exact thickness of the pamphlet, and make allowance for this thickness in the inner margin or back. When it can be done, put marks between the second and third pages of a cover, indicating the thickness of the book, as a guide to the coverer.

Bodkins.—The utility of these little instruments is undoubted, and, in the hand of a capable workman, effect a great saving of time, though manufacturers perpetrate one great fault in making them, and that is they are too long and badly finished ; seventy-five per cent. break after a very brief use. We hear and read of different opinions concerning bodkins, but at least compositors know how to manipulate them, and complaints of injury to the type by the bodkin are rare. Our American friends, on the contrary, as a rule, denounce the bodkin in terms that would amount to its being a friend of the type-founder. Why should practical men advocate its extinction ?

Arrangement of Case Racks.—As far as possible keep them so as to have a series of faces in the same rack. By that means a larger cap will often suggest itself to make a more effective display line.

Method of Laying New Type.—The careful laying of new type into case is a work of greater importance than most printers believe. It is too often intrusted to apprentices, without instruction as to the method of doing it, the result being often seen in cases of pie and a large proportion of battered letters. The following plan is recommended on laying a new book fount, and on no account should a learner be allowed to undertake the work before he can distinguish from each other the letters b q, u n, d p, as well as the small capitals c, o, s, v, w, x, z, from the same letters in the lower case. In regard to these small capital letters, however, some type-founders adopt the plan of giving them an extra nick to distinguish them from the lower-case. Unwrap carefully the page received from the founder, and, laying it on a galley, it may be thoroughly soaked with thin soap-water, to prevent the types adhering after they have been set up and worked off; then, with a stout rule or reglet, lift up as many lines as will make about an inch in thickness, and placing the rule on one side of the bottom of the proper box, slide off the lines gently, taking care not to rub the face of the letter against the side of the box. Proceed thus with successive lines till the box is filled. Careless compositors are prone to huddle new type together, and, grasping them in handfuls, plunge them pell mell into the box, and then roughly jostle them about to get more in. The type left over should be kept standing in regular order until the case needs replenishment. A fount of 500 lbs. of pica may have, say, five pairs of cases allotted to it ; the same amount of nonpareil from eight to ten pairs of cases.

Setting Short Measures.—Say the measure is two ems brevier. Make up the stick to ten ems, drop in two four-em quads of its own body in the end of the stick farthest from you, justify up to them, and go on filling the stick ; when it is full, empty it and remove the quads.

A Good Suggestion.—Always pick up a type, lead, rule, or quoin, at the time it is dropped. This is not only a saving of material, but it engenders a habit of carefulness and economy. Moreover, the stooping and bending of the body is often a relief, especially after standing erect for some time.

Cleaning Formes with Steam.—Steam has the advantage of rapidly boiling the oil of the ink, which condenses ; it gets rid of all dirt, and leaves the types perfectly clean. Further, types cleaned by this means always look new, and that the oxidation produced by potash, which is so injurious to the skin, is avoided. Let the forme be subject to the jet of steam for two minutes. The heat will dry the types instantaneously, and much facilitate distribution. As no brushes and potash are required, the expense of fixing up the piping is very soon saved. The steam must be drawn direct from the boiler, as waste steam is not hot enough.

Washing Formes.—Formes sent down to machine ought not to be wet too much with lye or with water, otherwise it becomes necessary to dry them before working, which takes time and often much trouble. The wet works up little by little to the face of the letter, and then the forme becomes unworkable. It has often to be taken off the coffin, the feet of the types have to be thoroughly dried, then some sheets of unsized paper have to be placed under the forme ; it has also to be unlocked, shaken, locked up again, the sheets removed with the moisture they have imbibed, and then it is to be hoped the forme will be workable. If not, there is nothing to be done but to lift it and dry it by heat.

Lye is generally used for washing formes which do not contain wood blocks ; turpentine where woodcuts or wood-letters are to be found in them. The bristles of the lye-brush should be longer than those of the turpentine-brush, and, in order to preserve it, each brush should be properly washed with water after using, and shaken and stood up to dry. If this is not done the brush will last but a short time.

There is no good in taking up with the brush a large quantity of lye or turps, and to shed it at once. Yet this is too commonly done, regardless of waste. In order to wash a forme well the brush should be passed lightly over all the pages, in order to wet them uniformly. Then they should be rubbed round and round, and finally lengthwise and crosswise. Leaning on the brush not only wears away the bristles, but sometimes injures the face of the types. It is a bad practice.

Barbed Spaces.—A foreign journal draws attention to a useful little invention just invented, and intended to facilitate the keeping of type, ornaments, and so on, in blank cases. It consists of slips of white metal, with projecting spurs on each side, so that they can be firmly driven into the wood and fixed there. In this way the cases can be subdivided with the help of wooden sticks into a number of compartments with firm sides. The invention has claimed for it the following, which are named to us :—The stamps no longer fall down : pie is avoided ; the stock of type can be controlled and the letters examined ; the type is protected from injury and does not stick fast ; the process of setting up and distributing the type is considerably facilitated, valuable spacing material, such as quads, clumps, furniture, etc., is saved ; the cases can be kept in order for an indefinite period ; compartments of any required size can be formed instantaneously in no other way ; ornaments, borders, rules, polytypes, electros, and similar things, can be preserved in the same way by the application of these spaces.

Buying Type.—When buying a fount of type try the metal by cutting it with a knife. You will soon discern the difference between good and bad metal. Above all, do not be too anxious to buy cheaply. It is not always that a fount of type sold is worth the money paid for it. In buying job founts it is a profitable investment to purchase the whole of the series. Never ask a founder to divide a fount. It is often more economical to buy double founts, and thus avoid picking and turning for sorts. Too small founts are entirely useless. Quite a mistaken notion is it that cash not spent in new type is money saved. Find the man who has this mistake in his head, and who allows it to rule his business conduct, and you will most probably find one who is not troubled with a flourishing business. The reason is not far to seek. Although a single evil may be borne by customers, who can stand bad type, bad ink, and bad paper—especially when by going a few yards further good type, ink, and paper can be relied upon ? Some printers work their type as long as there is any of the stamp left, and then would like to turn it round and print the fount from the other end! Another mistake is made when it is supposed that an ornamental job is not a profitable one, simply because it takes time in composition. Of course we say this of a properly appointed office. Our contention is that a good job can be done quicker in such a place than any kind of work can be turned out from a badly appointed office. In other words, it pays to keep pace with the times; and the advice is sound which recommends the printer to let nothing but the length of his purse restrain him in laying in new material.

Compositors' Requisites.—Every jobbing hand should possess a gauge, one up to 100 ems, and a sheet containing the sizes in inches of the various folds of paper, so that on receiving instructions to set a job he would not be at a loss for some idea of the size. It is in the seemingly unimportant items where valuable time is lost.

Standing Formes.—In almost every office of any importance there are hundreds, and perhaps thousands, of formes standing. Some of them, probably, are only used once a year, nevertheless they are allowed to stand there until wanted again. The method is considerably on the increase, and overseers would be acting wisely to thoroughly overhaul the list of standing formes. How often may the men be noticed going from one drawer to another for quoins, etc., which, by a judicious working of the standing formes, would release enough material in the shape of quoins, furniture, side and footsticks, and chases, to supply a good many men for days.

A Suggestion for Spaces.—Keep your middle and thin spaces separate. It is no uncommon occurrence to wade through a box of mixed spaces to get half a dozen thins, whereas, if kept separate, they could be obtained at once. It is just as ridiculous to mix them as it would be to mix the lower case t's and a's, and expect to know the difference by their thickness in using them. If the case maker would arrange the case with another box for thin spaces—say by taking about one-third of the upper end of the lower-case e box, on the nearest side to the middle space box—the practice of mixing thin and middle spaces would soon die out.

Spacing of Matter.—Solid works should be rather closely spaced than driven out. In leaded works, with a choice of wide or close spacing, always adopt the wider. Do not divide words at end of lines ; this should only be tolerated in narrow measures or exceptionally large types.

Self-spacing Type.—The following illustration explains the system recently introduced by an American foundry :—

One unit.—Space   1
Two units.—Space, f i j l . , : ; ' - ! ɪ ' '.   Total, ...   14
Three units.—Quad, c e g r s t z ?) † ‡ §] ‖ ¶ ɪ c ȷ s
    ᴛ ᴢ – °. Total, ...   25
Four units.—Quad, a b d h k n o p q u v x y fi fl ff $ £
    1 2 3 4 5 6 7 8 9 0 C J S Z ᴀ ʙ ᴅ ᴇ ꜰ ɢ ʜ ᴋ ʟ ɴ
    ᴏ ᴘ Q ʀ ᴜ ᴠ x ʏ & ¼ ½ ¾ ⅓ ⅔ ⅛ ⅜ ⅝ ⅞ – J ⸜ ⸓ |.
    Total, ...   67
Five units.—w æ œ A B D E F G L N O P Q R T U V
    Y & ᴍ ᴡ ᴁ ᴔ ℔. Total, ...   25
Six units.—Quad, m ffi ffl H K M X — ... ℞. Total, ...   11
Seven units.—W Æ Œ @.   Total, ...   4
Twelve units.—Quad, ☞ ☜ — ...... Total, ...   5

Grand total to the fount, ... 152

Spacing.—Some text-books tell us a thick space in the case of Roman lower-case, and an em quad generally in the case of capitals.   Clearly these are misleading.   Are we to space a condensed Roman lower-case letter as wide as one of bolder proportions, or a line of small-faced caps as wide as a line of large-faced ones?   Certainly not.   Then what rule is there which will apply equally well to all faces.   The proper space to use between words is undoubtedly that which most nearly approaches the average thickness (width) of the fount being used.   Practically the a, e, o, or u, of any fount, whether the line is in the lower case or capitals, are sufficiently near the average for our purpose, and as one of these letters always appears in every line, the proper space is at once apparent. This rule will hold good in all cases, because the thickness of the letter being taken it matters not on what body it is cast.

The proper or normal space being ascertained, the variations from it should be as small as possible. The compositor, however, should bear in mind the following points, which often render an increase or decrease of space desirable, owing to the amount of white near the line. Letters which are very condensed may be spaced a little wider than the rule gives. Letters which are much expanded may be spaced a little closer than the rule gives. When letters of a word are spaced out, the double of the space between the letters should be added to the space between the words.

Casting Leads.—The casting of leads is by no means easy, especially if long ones are required. There is a good deal of knack in the operation. The mould must be kept hot and smoked over a flame from time to time, or, better, rubbed over with a thin solution of jeweller's rouge in water. This greatly assists the metal to fill the mould.

Weight of Leads Required for a Job.—Multiply the number of lines in a page by the number of pages to be leaded, and divide the product by the number of leads of the measure required that go to the lb. (see separate table, p. 30).

EXAMPLE : I have to lead (8-to-pica) 24 pages of matter set to 21 ems pica, there being 35 lines to the page. How many pounds of leads shall I want ? In the table I find 54 8-to-pica leads, 21 ems long, go to the lb. Therefore I divide 35 × 24 by 54 and get 15 lbs. 10 oz. *Ans.*

NOTE. I should order 20 lbs., cut to the right measure, to be sure of having enough.

Proportions of English and French Type Metals.—English types are made of metal composed of 55 parts of lead, 22.7 of antimony, and 22.3 of tin ; or 61.3 of lead, 18.5 of antimony, and 20.7 of tin ; or, again, 69.2 of lead, 19.5 of antimony, 9.1 of tin, and 1.7 of copper. French metal is composed of 55 parts of lead, 30 of antimony, and 15 of tin. Besley's metal was composed of 100 parts of lead, 30 of antimony, 20 of tin, 8 of nickel, 5 of cobalt, 8 of copper, 2 of bismuth.

Job Fount Schemes.

Founts of jobbing type are now being sold on the Aa basis instead of by weight. The following schemes, comprising a large and a small fount respectively, will therefore be found useful by some printers.

4A 20a FOUNT.						36A 70a FOUNT.					
A	4	a	20	1	5	A	36	a	70	1	16
B	2	b	8	2	5	B	16	b	24	2	12
C	3	c	12	3	4	C	22	c	34	3	12
D	3	d	16	4	4	D	20	d	36	4	12
E	5	e	35	5	4	E	42	e	92	5	12
F	2	f	8	6	4	F	18	f	24	6	12
G	2	g	8	7	4	G	18	g	24	7	12
H	3	h	16	8	5	H	22	h	44	8	12
I	4	i	20	9	4	I	36	i	70	9	12
J	2	j	5	0	6	J	10	j	16	0	16
K	2	k	6	§	4	K	10	k	12	§	10
L	3	l	16	£	1	L	22	l	44	£	5
M	3	m	12	&	5	M	20	m	32	&	10
N	4	n	20	fi	—	N	36	n	70	fi	7
O	4	o	20	ff	—	O	36	o	70	ff	7
P	3	p	12	fl	—	P	20	p	26	fl	5
Q	2	q	5	ffi	—	Q	8	q	10	ffi	4
R	4	r	20	ffl	—	R	36	r	70	ffl	4
S	4	s	20	,	20	S	36	s	70	,	50
T	4	t	20	;	4	T	36	t	70	;	14
U	2	u	12	:	4	U	20	u	34	:	12
V	2	v	8	.	20	V	10	v	12	.	50
W	2	w	8	-	8	W	12	w	20	-	18
X	2	x	4	'	8	X	8	x	10	'	24
Y	2	y	8	!	4	Y	12	y	24	!	14
Z	2	z	4	?	3	Z	8	z	10	?	12
Æ	1	æ	1			Æ	3	æ	4		
Œ	1	œ	1			Œ	3	œ	4		

Jobs on Hand-made Paper.—Small jobs should always have the deckle or rough edge on the tail ; and if a side deckle also, should be so printed as to have the rough edge on the right of a *recto* page and the left of a *verso* page as far as possible, if the paper will allow of it.

Hints on Ordering Sorts.—A typefounder has published a table containing a rough estimate (taking Brevier as a standard) of the amount which the respective boxes of the regular full-sized lower case will contain ; the first two columns give the letters and weight only, the last two columns the letters and number that will weigh a pound.

Letters.	Weight to Box.	Letters.	No. Letters to lb.
a c d i s m n h o u t r	2 lbs.	a b d g h k n o p q u v x y z	582
f b l v g y p w	15 oz.	c e r s t	682
k j z x q and figures	6 oz.	m	398
e	3 lbs.	f i j l	850
Caps	5 oz.	Periods and Commas	1400

Composing Room Tools.—It is astonishing to go into some large printing establishments and find therein a woful disregard of modern improvements. For instance, in most houses there is generally a " ship" exclusively for small jobbing, but here you will have the imposing surface, mallet, planer, and shooter just as large as a " ship" doing heavier work. Probably some will say that a large imposing surface is useful for an emergency. This may be so, but there is no earthly reason to use a mallet and planer of the ordinary size on delicate cards, etc. Overseers of composing rooms should issue to their clickers mallets and planers about one-third the size. They are much handier, just as effective, and the comp. can better feel whether he is injuring the type or not with proper tools.

Setting-up and Working Numbers.—Suppose five sets of numbers from 1 to 100 are required, first set up the ten digits in a column and print 240 copies to the right of the centre of the sheet. Then shift the sheet on the press so as to print ten copies of the same column side by side with the first ; this makes 11, 22, 33, etc. Now remove the cipher from the bottom and place it at the top, so that the column reads 01, 12, 23, etc. The 9 is now put at the head of the column, causing it to read 91, 02, 13, etc. One by one the figures are transposed from the bottom to the top, the last column reading 21, 32, 43, etc., of which, as with the others, print ten copies. Then set up ten 1's, and print them on five of each of the forms already done, except the first, and also changing the lower 1 to a 2 on the column ending with 00. This, it will be seen, gives us all the numbers from 1 to 200. The operation is illustrated by the following table, showing the successive printings. Bear in mind that the left-hand column (1 to 0) is printed on all at the first running through of the press :—

1	11	01	91	81	71	61	51	41	31	21
2	22	12	02	92	82	72	62	52	42	32
3	33	23	13	03	93	83	73	63	53	43
4	44	34	24	14	04	94	84	74	64	54
5	55	45	35	25	15	05	95	85	75	65
6	66	56	46	36	26	16	06	96	86	76
7	77	67	57	47	37	27	17	07	97	87
8	88	78	68	58	48	38	28	18	08	98
9	99	89	79	69	59	49	39	29	19	09
0	00	90	80	70	60	50	40	30	20	10

It will be seen that but twelve formes (including the 1's) will be required for this work, while much less time will be spent in making changes than would be occupied in endeavouring to print after the old method. Of course if the numbers and changes run above 200 it will only require a greater number to be printed on each form, with the corresponding addition of higher numbers for the hundreds.

A Few Don'ts for Comps.—Don't use leads in quoining up a forme.

Don't use two quoins together in fitting quoins ; the construction and shape will not permit of it.

Don't neglect to pick up types when you drop them on the floor.

Don't place letters upside down for " turns," but reverse the nicks.

Don't pick good matter for sorts.

Don't spike lines with the bodkin when badly justified and won't lift.

Don't lock up formes as if you were at a blacksmith's forge.

Don't forget in planing down that type is *malleable.*

Don't space unequally or divide words at the end of lines if it can be avoided.

Don't shirk any method of improving your work.

To Preserve Wood-Letter and Cuts.—To prevent warping in blocks and wood-letter used in large bills, they should be placed in a zinc basin, provided with an air-tight lid, and then thoroughly saturated with paraffin oil ; after being left thus for about four days they should be wiped with a clean, dry rag. Prepared in this way, when new, wood-letter is stated to resist the effects of lye, petroleum, turpentine, and atmospheric changes.

Another Method of Preserving Wood Letter.—It is generally believed that oiling the face of new wood letters, or even of woodcuts, will make them take the ink better, besides giving them greater durability. But in the first case the result has not always been favourable. A foreign machine-minder has found a means of obtaining the desired effect, without risk of making the types or cuts too greasy. He pours some oil on the imposing stone, or, better still, on an iron slab, and spreads it with the finger on a space the size of the wood letter or cut to be oiled. Putting the cut then, face upwards, on the oiled spot, in a very short time the oil is entirely absorbed, and the letter or cut is permeated so entirely as to protect the finest lines of its face.

Repairing Battered Wood Type.—The last office I worked in (writes a correspondent of the " Inland Printer ") was stocked with battered wood type, of course caused by careless handling on the press ; broken tapes, dirt, and an occasional falling out of one of the feed guides on to the forme while in motion had caused the trouble, and it was impossible to do good work with such an outfit. I tried filling up the depressions with sawdust and glue, beeswax, etc., but the result was not satisfactory. I determined to conquer the difficulty, and, after devoting considerable thought to the matter, I mixed some warm glue with Spanish whiting, and, after cleaning out the depressions well, and in some instances deepening them in order to give the preparation a good chance to hold, I plastered the defects over with the mixture while warm. I put sufficient on to thoroughly fill all depressions, not being careful to get a smooth surface. After it became hard I filed it down close to the letter, avoiding scratching the even surface of the letter, and then treated it to a good rubbing with an oil-stone, using oil, and the result was a polished surface as good, if not superior, to the wood itself; and, as I rubbed down the plaster even with the surface of the letter, the printing failed to show any defects whatever. Even the planer did not damage it, and I felt much elated in overcoming the difficulty.

Copper-faced Wood Type.—Attention is called to an invention made by a Parisian, which, it claims, will cause quite a revolution in the typographic world. In short, he has been experimenting with the galvanic process and wooden type, and has succeeded in induing the upper surface with a coating of copper. The importance of the invention can hardly be over-estimated, especially to those who use a great deal of wood letter and know its perishable nature. In order to produce what the inventor calls galvanized wooden type, the letters are placed in a galvano-plastic bath, and receive a coating of copper. The letter, while preserving nearly the lightness of wood, is as strong as metal, the copper coating rendering its form unchangeable, and preserving the wood from exterior influences and other risks which result from the ordinary manipulations to which type is subjected. The covering of the upper surface with galvanized copper has also the effect of preserving the delicate serifs of the letters, which are as strong as type-metal. Although the inventor at present only employs his process for broadside type, he purposes so far to improve his method as to permit of its application to heraldic engraving, to escutcheons, arms, etc. The possibilities of its application, indeed, are indefinite ; for instance, it may give wood engraving a fresh chance, as it may be possible, under this process, to use "cuts" for long numbers where now electrotyping has to be resorted to, much to the detriment of the artistic finish of the work.

How to Treat Wood Type.—To prevent warping, all very large wood type should be set up on the edge when put away, so that both sides may be equally exposed to the air. In cleaning it, neither lye nor water should be employed under any circumstances. Turpentine, paraffin, benzine, or kerosene oil may be used ; but turpentine and paraffin are the best. Procure a small shallow pan ; lay the forme flat on the board ; pour about six table-spoonsful of turpentine into the pan ; touch the face of the brush to the turpentine, and pass it quickly over the forme before it evaporates. Six or eight spoonsful of fluid will be found sufficient to clean a large forme, if thus used.

Keeping Cases Clean.—Dust is a great foe to types, as well as to the comfort of the compositors. No sane person would think of throwing sand upon a forme to be ground into the types. Yet the damage is proportionately as great when cases and types in constant use are allowed to accumulate dust to be ground against the faces of the types by the shaking of them up, the jarring consequent upon the con-tinual touching of the fingers in the case in composition, and the repeated friction that comes from use. It is a matter of economy to have cases blown out at least once a week, and piece hands will make money by keeping their cases free of dust by the increased amount of work they will be enabled to do.

Sticky Type.—It is said that types, especially new ones which have been papered and put away for a long time, and which consequently stick badly, may easily be separated by placing them on the stone and pouring a little glycerine upon them, leaving them to stand there over night. The glycerine may be washed away with warm water, when the types will be found ready for distribution.

Another Remedy for Sticky Type.—The type in long-standing jobs often becomes firmly cemented together and incrusted with dirt, and resists strongly all ordinary attempts to separate it. Pour lard oil over the face of the type, rubbing it in thoroughly with a soft rag; then wash in strong lye with a brush and cleanse with hot water, and every type will be found loose and clean.

Another Remedy for Sticky Type.—A printer writes that he has experienced great annoyance in this line. After trying every plan suggested, with but little relief, as an experiment the foreman lifted a column of matter to a galley, and, after slightly locking it, took the benzine can and ejected benzine upon the face of the matter, then with fingers and thumbs worked the type back and forth slightly, but thoroughly. After standing a little while the matter was thoroughly saturated, and the caking of the type was found to be cured. If this experience is worth anything to the sore-fingered and discouraged distributor of new type, the object of this paragraph is accomplished.

Emergency Wood Letters.—A few sticks of elastic glue are useful in a jobbing office. Very often an extra wood letter is wanted for a line, above the number supplied in the fount. This can generally be met by a little alteration or addition to another. Thus an A formed from V and *vice versâ*, B from P, H from two II's, M from W, and by cutting L from T, etc., and in case the original letter is required it is but the work of a minute to either cut away the glue or to add it if the letter has been cut. A gas-jet and a pocket-knife is all that is needed. Work with the blade of knife heated.

First Use of the Setting-Rule.—This useful little implement was quite unknown to the early printers, and up to the time of its first adoption the lines of type (except in the case of the larger founts) varied in length like the lines of the manuscript, because the compositor was unable, without frequently breaking the line, to shift the words in order to increase or decrease the normal space between them. When setting-rules were devised, it so facilitated this operation, and, by making all the lines of an even length, so improved the symmetrical appearance of the pages, that no printer, after once trying it, returned to the old plan. In 1467 Ulric Zell, of Cologne, was unacquainted with this improvement, but as, out of the great number of works which issued from his press, it is a rarity to find lines of an uneven length, it is safe to conclude that he adopted it about 1468-69. But Meinsion, at Bruges, did not use it till 1478, ten years later, while it took nearly ten years more to cross the Channel to Westminster, where Caxton adopted it in 1490.

Twisting Brass Rule.—Work and designing in brass rule are now the most popular means of ornamentation, and a few practical hints may be helpful. We ("Inland Printer") soon learned to bend leads by heating them; but it was years before we learned that brass rules, even as heavy as nonpareil, could be easily bent after heating and allowing to cool. The rule must not get too hot, or it will melt; keep watch on it and remove it from the fire before reaching white heat. Circles, ovals, curves, and flourishes may thus be easily made at pleasure; and even letters for initials, very neat and unique in appearance, may be produced with only a file, a vice, and a hammer for tools. For most rule-work ten-to-pica rule will be found cheaper and easier managed than the heavier rules. When it is desirable to bend the corners of a rule up or down, time and trouble in justifying may be saved by cutting the rule parallel with the face, and just above the top of leads used, the desired space, and bending it in any direction desired. This is recommended for short bends only, though with heavier rules it will work for a longer space. A pair of pincers will be found very useful in doing rule-work.

c

Remember, that a good artistic worker in rule ornamentation can make a success, and command good prices, while a poor workman cannot do as well as a plain workman—that is, he cannot obtain as good prices in proportion to time spent as the plain workman. But we presume only those who have tact. genius, and a love for this style of ornamentation, will make a true success of it ; and we write for the multitude who may wish or need to do something in this line, either for pleasure or profit. To do successful work it is necessary to have good material and good tools to work with, and then after the design is finished it is essential that good ink be used, and great care be taken in presswork to bring out the effect in the printing.

Measures for Bookwork.

Size.	The Page of Type should measure in Pica Ems.		Size.	The Page of Type should measure in Pica Ems.	
	Length.	Width.		Length.	Width.
Foolscap.			*Demy.*		
4to. ...	41	30	4to. ...	54	42
8vo....	32	18	8vo....	42	24
12mo.	28	15	12mo.	36	19
16mo.	19	15	16mo.	26	20
			32mo.	21	12
Crown.			*Royal.*		
4to. ...	48	34	4to. ...	64	48
8vo....	36	21	8vo....	48	27
12mo.	32	16	12mo.	40	21
16mo.	23	16	16mo.	29	21
			32mo.	24	14

Emergency Tints.—It is said by a writer that in the absence of the proper colours for a job wanted quickly, he has found the artists' oil colours (retailed at stationers' shops in collapsible tubes) to answer excellently mixed with transparent tinting ink. He says "they work up well, print clean, and look fresh and bright."

Lengths and Widths of Pages for Ordinary Bookwork,

With the number of Ens contained in each page, from Pica to Nonpareil inclusive.

Length.	Width.	Size.	Pica.	Small Pica.	L. Primer.	Bourgeois.	Brevier.	Minion.	Nonpareil.
Picas.	F'cap.		Ens.	Ens.	Ens.	Ens.	Ens.	Ens.	Ens.
41	30	4to.	2460	3243	3825	4988	5796	6900	9840
32	18	8vo.	1152	1517	1800	2346	2695	3180	4608
28	15	12mo.	840	1088	1295	1680	1978	2350	3360
19	15	16mo.	570	748	888	1134	1334	1600	2280
21	12	18mo.	504	672	780	1020	1184	1400	2016
		Crown.							
48	34	4to.	3264	4290	5040	6693	7696	9120	13056
36	21	8vo.	1512	1968	2340	3060	3520	4200	6048
32	16	12mo.	1024	1369	1600	2116	2401	2809	4096
23	16	16mo.	736	962	1160	1518	1715	2014	2944
23	15	18mo.	690	884	1073	1419	1610	1900	2760
		Demy.							
54	42	4to.	4536	6014	6968	9240	10707	12600	18144
42	24	8vo.	2016	2640	3120	4080	4810	5600	8064
36	19	12mo.	1368	1804	2115	2754	3190	3780	5472
26	20	16mo.	1040	1380	1600	2109	2440	2838	4160
28	16	18mo.	896	1184	1400	1840	2107	2491	3584
21	12	32mo.	504	672	780	1020	1184	1400	2016
		Royal.							
64	48	4to.	6144	8140	9401	12604	14652	17120	24576
48	27	8vo.	2592	3410	4020	5313	6142	7290	10368
40	21	12mo.	1680	2208	2600	3420	3968	4690	6720
29	21	16mo.	1218	1584	1872	2520	2880	3430	4872
32	18	18mo.	1153	1517	1800	2346	2695	3180	4608
24	14	32mo.	672	896	1050	1360	1591	1880	2688

The above figures are subject to the slight variations in founts from different foundries.

Mitred Rule.—In examining work which has been submitted as a specimen of fine display printing, the practical and experienced critic too often finds what to him is a terrible eyesore, and one which is simply the result of carelessness or indifference, and therefore the less excusable. The allusion is to the absence of mitred corners in brass rule borders. The greatest difficulty is of course experienced in making rule join which has not been mitred, but there are plenty of firms who would lock-mitre the rule for a few pence per set of four corners, or, if such firms are inaccessible, a mitreing machine can be obtained at a cost of a very few pounds.

A Square Inch of Type.—It is not always, when asked for an estimate, that the printer has at hand the means of making it. He has mislaid his graduated scale, or the matter to be cast off is in awkward batches. A ready method, however, is suggested. It has been calculated that a square inch of pica contains 36 ems; of small pica, 49 ems; long primer, 56 ems; brevier, 86 ems; minion, 100 ems; nonpareil, 144 ems; and ruby, 196 ems. Any fractions in the calculations are in favour of the printer.

A Hint for Galley-Proofs.—See that they all are upon full-sized slips, no matter whether one galley is a short one or not. If submitted on a short slip, this is the one which is generally lost.

Useful Hint for Pulling Galley-Proofs.—Many comps. have a knack of bringing up the spaces when proofing a galley. This can be overcome, even when using the keenest roller, by rolling the galley diagonally, and giving the roller a gentle twist in passing over the type.

Clearing away Leads.—In distributing leave your leads and slugs to the last, and make one job of it. Take a long galley, jog up your leads as they come on the galley, then pick out the longest, next longest, and continue until you reach the smallest, packing the assorted leads at one end of the galley. Carry the galley to lead rack, and in a few minutes the job will be done, while if done by the haphazard plan which nine out of ten compositors use, much time would be lost.

Receptacle for Battered Letters.—Every compositor should keep by him, convenient to his right hand, a receptacle for battered letters and wrong founts, and on no account should he throw either into the quad-box, or into some spare box in the upper case. Where the latter plan is adopted, very frequently it turns out that comparatively scarce letters are thrown into it; whereas were they placed as here suggested, and distributed, say, once a week, much untidiness would be prevented, and the type would all the sooner be brought into use again. There is as much type carelessly hidden away in the quad-boxes of some printing offices as would fit up a small newspaper and jobbing office combined.

Comparative Sizes of Types.—The following are the average number of ems that go to the foot. The types of some founders vary slightly in the measurement of their bodies.

Pica	72 ems	Minion... ... ... 122 ems
Small Pica	83 ,,	Emerald ... 138.5 ,,
Long Primer	89 ,,	Nonpareil ... ... 144 ,,
Bourgeois ...	102.5 ,,	Ruby ... ... 166 ,,
Brevier	111 ,,	Pearl ... ... 178 ,,

EXAMPLE of use of the table :—A book set in Small Pica 21 ems (pica) wide, 36 lines to a page, leaded 8-to-pica, occupies 600 pages. If the same book should be set in Brevier solid, 18 ems (pica) wide, 45 lines to the page, how many pages will it occupy?

The proportions are as follows :—111 : 83, 45 : 36, 18 : 21. The fact that one book is solid and the other leaded need not be taken into account, as the difference is shown in the number of lines to the page. Therefore, the number of pages required is $500 \times 83 \times 36 \times 21$ divided by $111 \times 45 \times 18 = 349$ pages. *Ans.*

NOTE. The faces should be of the same series; that is, proportional. If the Small Pica were a condensed letter and the Brevier an extended, the number of pages would be correspondingly enlarged and *vice versâ.*

Relative Sizes of Type, from Pica to Pearl.

Pica.	Small Pica.	L. Primer.	Bourgeois.	Brevier.	Minion.	Nonpareil.	Ruby.	Pearl.
6	7	7½	8½	9	10	12	13½	15
7	8	8½	10	10½	11½	14	16	17½
8	9	10	11½	12½	13½	16	18½	20
9	10½	11	12½	14	15	18	20½	22½
10	11½	12½	14	15½	17	20	23	25
11	12½	14	15½	17	18½	22	25½	28
12	14	15	17	18½	20½	24	27½	30
13	15	16½	18½	20	22	26	30	33
14	16	17½	20	21½	23½	28	32½	35
15	17½	19	21½	23	25½	30	34½	38
16	18½	20	23	25	27	32	37	40
17	19½	21½	24	26½	28½	34	39½	43
18	21	22½	25½	28	30½	36	41½	45
19	22	24	27	29½	32	38	44	48
20	23	25	28½	31	34	40	46	50
21	24	26½	30	32½	35½	42	48½	53
22	25½	27½	31½	34	37	44	51	55
23	26½	29	32½	35½	39	46	53	58
24	27½	30	34	37	40½	48	55½	60
25	29	31½	35½	38½	42	50	58	63
26	30	32½	37	40	44	52	60	65
27	31	34	38½	42	45½	54	62½	68
28	32½	35	40	43½	47½	56	65	70
29	33½	36½	41	45	49	58	67	73
30	34½	38	42½	46½	50½	60	69½	75

Scarce Founts.—It is a good plan, when putting away standing formes for any length of time, to "lift" the lines set up in type of which there is likely to be a scarcity, and re-lock up.

The Origin of Italic.—The form of Roman now known as Italic was originally called Aldine. The first volume printed in this character had the capitals with their stems upright like those of the current round hand. These first editions were the works of Virgil, printed by Aldus Pius Manutius, in 1512, and it is known that this celebrated printer made use of a manuscript text entirely copied by Francesco Petrarca. Thus, it is said, that Manutius desiring to pay public and reverent homage to the author of the Canzoni, appropriately wished a hanging character cut in imitation of his writing, intrusting the design and the cutting to a skilled artist, one Francesco da Bologna. But the fashion of these editions in cursive italic type lasted only a short time, having been imitated by foreign printers in a careless and illegible manner. The cursive character was at that time known both in Italy and outside of the country under the name of Aldine, but later the title of cursive was given to it from the writing of the Roman Chancellery, called cursiveti seu cancellarii; a title which in Italy has superseded every other.

Accented Letters may be easily and effectually made by the shoulder of the letter being cut off, say, one-sixth of an inch, and, in the case of a diæresis accent, soldering on the top part of a colon with a small blowpipe ; the letter afterwards being filed and dressed. Other peculiar and accented letters can be made equally well : a little thought and ingenuity will overcome many obstacles.

Mallet, Shooter, and Quoin.—No end of time is wasted in perpetually having to take these useful articles out of a drawer whenever they are wanted. The proper place should be at one end of the surface, and that divided off into two portions—one-third for the mallet and shooter, and two-thirds for the quoins ; by so doing they are always at hand. If there is any objection as to its being in the way, an arrangement might be effected where this shelf could be slid under the surface whilst any correction was being made. The whole side arrangement need not be more than six inches in width.

The Water Jug and Sponge.—One of the most disreputable combinations in a composing room is the water jug and its sponge. Generally it is a beer jug with its handle broken, or a beer can with the handle off, to hold the water, and for a sponge in hundreds of cases, a lump of paper. Such articles do not tend to raise the tone of the workmen, neither do they show the manager in a very good light, as he is responsible for every detail. A man to rule comps. must have the eye of a hawk, and an almost despotic will, otherwise he will find himself simply the manager by name. Comps. are extremely shrewd, and if an advantage can be taken, such as coming in late, leaving cases on frames, cutting leads and brass at will, etc., they are sure to do so, and the only one to blame is he who lets them perpetuate an error.

A Useful Tip.—All compositors appreciate the difficulty of cutting a strip of card the exact length of a measure, and the annoying result of getting it too long or too short. This trouble can be overcome by cutting the card a little short, and then cutting it in two in the centre diagonally. The card will then come out true at each end, will bear on the type at all points, and if it swells will spread inwards.

Lifts and Formes.—In large houses the composing and machine rooms represent the two extremes of the building, therefore the formes have to be sent down in the lift. No provision whatever is made in the lift for the protection of the type ; consequently, it is a frequent occurrence to see the comp. running downstairs to repair a battered forme. Just the same thing happens again when the formes are worked off; the man who takes them out of the lift lays them where he can find room, and if no one happens to be looking he is not particular whether the faces are together or not, and small blame to him, as such things ought not to be left at his discretion. A cheap and lasting way out of the difficulty would be to procure, say, fifty demy stout straw boards, and then give instructions that no forme is to be put anywhere without one of these boards. The remedy is simple, and does away with the use of many a quoin, which, after all, is liable to slip almost at any moment from a set of formes.

Display Founts.—Keep these types in double cases, *with spaces and quads*. Two founts of any one size may generally be kept thus, and there is then no need to go to a fresh case, perhaps in a different part of the office, to space out, or to get rid of the spaces in distributing.

Displaying of Titles.—Aim at simplicity, whether a crowded or open title-page. Do not use fancy types, and rarely a black letter. If the work is in modern or old-style type, use the same period of face, and do not mix the two kinds. Good round letters are preferable to any condensed fount, and if red lines are used in the title, it is best to balance the colour by having, say, a red line at top and again at bottom. In large pages, a line in the middle might also be introduced.

Picking Sorts from Standing Matter.—If this is absolutely necessary, never leave a forme with a single letter taken out without "turning" for it. It frequently happens that a standing job gets worked off from a forme which has been picked by some careless workman, whereas, had he done as suggested, the faulty line would have presented itself on the first impression being taken, when matters might have been remedied.

The Value of Good Chases.—A well-made true chase is absolutely necessary for good printing, and will save much worry, and in the long run much money. A roughly-made chase, welded in the corners, the lumpiness of its inside surfaces merely scratched down with a coarse file without reference to smoothness or squareness, is a very expensive article in a printing office, even though it cost nothing at all. It throws type off its feet, so that it looks badly in print and wears out rapidly. It is extremely liable to pie formes, and one pied forme costs more than two good formes. A poor chase just as it comes from the forge will cost less than one finished by the most perfect machinery; but a fine machine-finished chase will cost a great deal less than one finished by hand, and have a uniformity the hand-finish cannot approach. Like almost all kinds of printers' machinery, the best is the cheapest.

The Use of the Long ſ.—The old-fashioned and long ſ should be only used as an initial or medial to any word, and not as a final. The ligature letters containing the long ſ also should be adopted when they fall together, as the serif of the ſ would not allow the following ascending letter to close up to it without breaking off, as these letters will show : ſt, ſh, ſb, ſk. Double ſſ and ſſi are cast likewise in one piece, but in no instance must a long ſ actually finish a word; the letter attached to it, when a ligature, of course may do so.

Approximated Table of Type-Founders' Sizes.

	Caslon and Co.	V. and J. Figgins.	Sir C. Reed and Sons.	Miller and Richard.	Stephenson and Blake.
	ems to ft.	ems to ft.	ems to ft.	ems to ft.	ems to ft.
English	64	64	$64\frac{1}{2}$	64	$64\frac{1}{2}$
Pica	72	$71\frac{3}{4}$	$71\frac{7}{8}$	$71\frac{1}{2}$	$72\frac{1}{4}$
Small Pica	$82\frac{1}{4}$	83	$82\frac{1}{6}\frac{5}{6}$	83	83
Long Primer	89	90	$88\frac{7}{8}$	89	89
Bourgeois	$101\frac{2}{3}$	$101\frac{7}{16}$	102	$102\frac{1}{2}$	102
Brevier	111	$108\frac{1}{4}$	$110\frac{1}{3}$	111	$110\frac{2}{3}$
Minion	$121\frac{5}{16}$	122	122	122	$122\frac{1}{3}$

Ingenious Printing Device.—An invention, devised by a printers' engineer in London, accomplishes, it is said, a decided reduction in the cost of such printing surfaces as types and blocks, enabling them to be produced in a much shorter time than is at present practicable. A layer of composition of glue and glycerine or treacle, such as that used in the manufacture of printers' rollers, composes the printing surface, being mounted upon paper or straw board, then upon a wood or metal base, type high ; a rotary cutter is then used to cut out the required letters or design from the composition, the surrounding material being stripped off. The design may be in intaglio, or relief, as desired.

Bastard Types.—Many printers and even typefounders apply the term "bastard" to types without knowing its meaning. Small pica and bourgeois are no more "bastard" types than long primer or pica. A "bastard" fount is one where a face from one size is put upon a body of another, as minion upon brevier body or *vice versâ*. Small pica, bourgeois, etc., were first termed "irregular" bodies because they varied from an imaginary standard instituted by someone in the early history of printing—probably based upon the difference of a sixth of a pica between sizes, as between pica and long primer, long primer and brevier, etc. As founders now make their bodies and faces by numbers it is doubtful if any real "bastard" founts can be found, except upon some of the daily newspapers, which demand large faces upon small bodies.

Difference between Type Bodies.—The following list will show the degrees

Between Small Pica and Pica	8	to pica.
„ Long Primer and Small Pica .	16	„
„ Bourgeois and Long Primer ...	10	„
„ Brevier and Bourgeois	20	„
„ Minion and Brevier	16	„
„ Emerald and Minion	30	„
„ Nonpareil and Emerald.........	18	„
„ Ruby and Nonpareil	16	„
„ Pearl and Ruby..................	30	„
„ Diamond and Pearl	24	„

Stone Type.—It is announced that an attempt is being made to make printers' type of artificial stone. It is claimed that this stone can be readily moulded, and is cheap; that it is hard, yet sufficiently elastic to bear great pressure without injury, whilst the type moulded from it will readily take up, retain, and give off the ink. Of course it is much lighter than the ordinary type-metal. Another inventor has promised to produce for us galvanized large-sized wooden type,—the top of the letter only receiving a coating of copper by a galvano-plastic application.

What makes Durable Type-Metal?—Printers as a rule labour under the impression that type, to wear well, must be made of hard metal. This idea is now being combated as a fallacious one, and an article on this subject says that the deep-rooted notion that hard metal makes the most enduring type is a mistaken one. It boldly takes the ground that the claims for a hard metal are a delusion and a snare, arguing that the mistakenly-prized hardness is obtained by using a large proportion of antimony, the cheapest of metal next to lead; and that hard type, under the action of the planer and the press, and also in distribution, will suffer in the fine lines and serifs, these being easily broken. These grave objections can be readily obviated by making type of a metal that is not only hard but tough, and to accomplish this desideratum tin and copper, the most expensive metals in the alloy, should be freely used.

Type-Metal.—Joseph Moxon, who was the first writer upon the technique of type-founding, printing, etc. (1683), gave as his receipt for type-metal the following proportions: 25 lb. of metal lead to 3 lb. of iron and antimony melted together. In Germany about the middle of the last century, according to Smith, type-metal was a mixture of steel, iron, copper, brass, tin and lead. We read of a printer who cast types "after a peculiar manner, by cutting his punches in wood, and sinking them afterward into leaden matrices; yet the letters cast in them were deeper than the French generally are."

Borders.—To be agreeable these need not surround a page if it presents a more agreeable effect when run upon two sides only. Still, the idea of uniformity and perfect balancing of parts is followed so tenaciously by many printers that they seem to find it impossible to use borders and parts of designs decoratively. The Germans particularly adhere to the exact and ponderous style, and while their work is, in many instances, exceedingly beautiful, it is quite as frequently so heavy as to be displeasing. The happy medium between the two extremes of excessive ornament and meaningless and insufficient ornament is one which only practice and observation will teach.

Printing Music as it is Played.—An apparatus has been invented which, when placed in electric communication with a pianoforte or other keyboard instrument, prints the music as it is being played. The machine is driven by clockwork. When in motion a band of paper is drawn through the machine, and during its passage is ruled with the staff lines and passes under a number of small marking wheels, which correspond to the keys of the pianoforte, and it is here that the music is recorded. The notation in which the music is recorded is not the ordinary one. When a note is depressed on the keyboard, which is in connection with the recording machine, the corresponding wheel will begin to mark, and the length of the line made will be in proportion to the length of note. Thus, if a line a quarter of an inch in length represents a quaver, a line an eighth of an inch will represent a semiquaver. This system of notation is easily translated into the ordinary notation by anyone who understands the latter. The connection with the keyboard is obtained by a pin fixed under each note which dips into a small cup of mercury. This arrangement in no way interferes with the touch of the keyboard.

A New Method of Music Printing.—Set up the staves in brass rule and work them first, then by means of special music types set up the musical notes, rests, bars, and other signs. These are all set to gauge, and then the forme is printed off on the sheets already bearing the staves, the only requisite being accurate register. By this system music may be printed in several keys from the same forme of type, all that is needful being to remove one or more leads from the bottom of the page to the top, or *vice versâ*.

Wooden Furniture.

Double Broad *is* 8 ems pica wide.
Broad & Narrow 7   ,,   ,,
Double Narrow   6   ,,   .,
Broad   4   .,   ,,
Narrow   3   ,,   ,,

the smaller sizes come under the head of "reglet."

Preservation of Type-Cases.—To protect type-cases and boards against the influences of damp, German manufacturers of such are treating the different parts of them with hot oil, impregnating them thoroughly before putting them together. They will never warp after having undergone this treatment.

Number of Leads in a Pound.

Lengths	4 to Pica.	6 to Pica.	8 to Pica.	Lengths	4 to Pica.	6 to Pica.	8 to Pica.
4 ems	144	216	288	26 ems	22	33	44
5 ems	112	168	224	27 ems	21	31	42
6 ems	96	144	192	28 ems	20	30	40
7 ems	82	123	164	29 ems	20	30	40
8 ems	72	108	144	30 ems	19	29	38
9 ems	64	96	128	31 ems	19	28	38
10 ems	56	84	112	32 ems	18	27	36
11 ems	52	78	104	33 ems	17	26	34
12 ems	48	72	96	34 ems	17	25	34
13 ems	44	66	88	35 ems	16	24	32
14 ems	41	61	82	36 ems	16	24	32
15 ems	38	57	76	37 ems	15	23	30
16 ems	36	54	72	38 ems	15	22	30
17 ems	34	51	68	39 ems	15	22	30
18 ems	32	48	64	40 ems	14	21	28
19 ems	30	45	60	41 ems	14	21	28
20 ems	28	42	56	42 ems	14	21	28
21 ems	27	40	54	43 ems	13	20	26
22 ems	26	39	52	44 ems	13	19	26
23 ems	25	37	50	45 ems	13	19	26
24 ems	24	36	48	46 ems	12	18	24
25 ems	23	34	46	47 ems	12	18	24

The Preservation of Page-Cords.—These may be rendered very durable by putting for an hour in a solution of lime, drying, and subsequent immersion in tannin. After being taken out and once more dried, they are saturated with oil.

Zinc Rules.—Zinc will not stand atmospheric influences —the face will oxidize, and crumble away in time. The mixture of zinc, also, either with stereo plates or types, is a fatal error.

A New Method of Making Blocks for Tinted Patterns has been patented by a machine-minder at Berlin. He uses all sorts of textile matter, preferably linen or cotton damask, with woven-in flowers or other figures, and begins by putting them in a bath of tannin and spirits of wine for twelve hours. Then it is inserted in a frame which may be extended by means of screws and wedges equidistantly, to open somewhat the space between the threads; then the frame with the matter is put in a drying-chest, and exposed to a heat of 20°to 25° Celsius. An hour later the heat is raised to 50° C., and molten beeswax is spread on the textile matter, which is still left for a space of about nine hours and at a heat of 70° C. in the drying-chest. Then the third process is come to. A solution of resin of Damar in turpentine is spread over the textiles; after remaining again for two hours in the chest, that process is repeated, and the drying continued from two to three hours. Now it is ready for printing, and may either be stuck by a strong glue on a basis of type-metal, or nailed on a wooden block. Such plates may be prepared nearly to any size, print much easier than large plates of metal, and last for very long numbers. If desirable to show wording on them, it may be done by cutting the types out of strong paper and pasting them on the printing cylinder after having finished the making-ready. The power of the printing will then be stronger on those parts, and the parts of the paper met by the extra pressure will, of course, appear some shades darker than the other parts of the tint-plate.

Overrunning Type in Corrections.—Many compositors in correcting do not overrun the matter as they ought to do in the stick, but on the stone, and frequently hair-space or treble-space a line, in order to get in or drive out a word; when by overrunning a line or two forward or backward they might preserve uniformity.

Weight of Type Required for a Job.—Divide the area of the page expressed in pica ems by 128. The answer gives the number of lbs. weight in the page. 50 per cent. for small founts, and 30 to 40 per cent. for large founts, should be added to allow for sorts, &c.

Example: I have to set fifty pages of Brevier octavo, the size of the page in pica ems being 20 × 34. What fount of type should I order?

The area of each page is 20 × 34, equal to 680 ems pica ; divide by 128 and multiply by the number of pages, 50. The result is 266 nearly. Add 40 per cent. and the sum will be 372 lbs. *Ans.*

The Sizes of Types.

Double Pica is 2-line Small Pica.		$4\frac{1}{3}$ ems Gt. Primer is 1 inch.	
Paragon is „ Lg. Primer.		$5\frac{1}{2}$ „ English is „	
Gt. Primer is „ Bourgeois.		6 „ Pica is „	
English is „ Minion.		7 „ Small Pica is „	
Pica is „ Nonpareil.		$8\frac{2}{3}$ „ Lg. Primer is „	
Small Pica is „ Ruby.		$9\frac{1}{2}$ „ Brevier is „	
Lg. Primer is „ Pearl.		12 „ Nonpareil is „	
Bourgeois is „ Diamond.		$17\frac{1}{3}$ „ Diamond is „	
Brevier is „ Minnikin.			

A shilling standing edgewise is type high.

The Setting of Half-Measures.—Matter set to two measures can, with a little practice, be just as easily managed by scraping a line on the setting-rule with a bodkin, and spacing to the engraved line, as by using a clump or lead as a " jigger." To make this clear, we will suppose you have first column 12 ems and remainder 14—26 in all. Mark your rule 12 ems from the front, and, in setting, space out to line engraved, then complete the line (the remaining 14 ems) in the usual way. Of course, if the columns are to be divided by rules, it will be safer to adopt the old method of using a "jigger," but if not, the foregoing plan will be found quite satisfactory, especially after a little practice.

Rough Proofs.—It has been said, "Never show a rough proof to a foolish client." We would go much farther—never show a rough proof to *any* client. Printers lose far more than they have any idea by showing rough proofs. To begin with, the client is disgusted, and first impressions [the pun forced its way in uninvited] are everything. It is no use saying, "This is only a rough proof," because for any meaning it conveys you may as well say "Abracadabra folderiddlelol." A client has been known to take a rough proof and show it round amongst his friends as a specimen of so-and-so's printing. There are successful printers who at an early stage of their career grasped the force of what we advance. From the first they got out their proofs in a workmanlike manner on good paper, and great has been their reward.

Leads required to Justify with Type from Pearl to Great Primer inclusive.

PEARL.............. One four and one eight-to-pica.
RUBY .... One four and one six-to-pica.
NONPAREIL ...... Two fours ; three sixes; or four eights.
EMERALD......... One four, one six, and one eight.
MINION .... One four and two sixes.
BREVIER .... Two fours and one eight.
BOURGEOIS ...... Three eights and two sixes.
LONG PRIMER... Three fours ; or six eights.
SMALL PICA...... Two fours and two sixes.
PICA .... Four fours ; or six sixes.
ENGLISH .... Three fours and two sixes.
GREAT PRIMER . Four fours and two sixes.

The Care of Twisted Rule.—When this has been used for ornamentation, and is turned out in distribution, it should be placed in a conspicuous position for future use, and not planted or hid by the person who happened to "twist" it in the first instance. Where this "planting" practice is allowed it must necessarily result in a much greater amount of rule being bent than need be for the requirements of the office.

D

New Composing Stick and Rule Cutter.—A prac-

tical printer of Camden, New Jersey, has invented an ingenious improvement. This improvement consists of a gauge, designed for both stick and cutter, by which either can be instantly and accurately set to any desired number of ems, pica or nonpareil, without the use of leads, rules, or quads. The cutter has along one edge a row of grooves, a pica em apart, which engage the lugs on the gauge. The cut is a sheer cut from the front, preventing slipping, and the lengths absolutely accurate. The stick sets with a thumb-screw, reversing to half measure, and marks indicate the length set to. There is no necessity to gauge with rules ; on the other hand, the correctness of rules or leads may be tested with the stick.

The Composing-Stick was first introduced as a printer's

tool in 1480. Previous to this the method of composition was by taking the letters direct from the boxes, and placing them side by side in a coffin made of hard wood, with a stout bottom, and kept tight when completed by means of screws at the foot.

A New Composing-Stick has a movable arm which

comes at the beginning of the lines, is in two parts, and secured by two screws. By loosening the one which is nearest the matter it can be instantly set for a half or third measure, or for any number of ems, while the moment this ceases to be requisite it can be drawn back till it meets the other part, when it is again at the full measure without a second's loss.

Poster Founts.—Where bill type is kept in open cases

or trays, difficulty is often found in stowing away large founts or expanded ones, as they take more than one case. If the racks are arranged side by side, then the heavier founts can be placed in trays in two or more racks parallel with each other, and each drawn out while getting or distributing a line, which is much better than having them underneath each other, for obvious reasons.

Tables of Signatures and Folios.

No.	Sigs.	Folio.	4to.	8vo.	½ Sh. 8vo.	12mo.	½ Sh. 12mo.	16mo.	18mo.	
1	B	1	1	1	1	1	1	1	B	1
2	C	5	9	17	9	25	13	33	C	37
3	D	9	17	33	17	49	25	65	D	73
4	E	13	25	49	25	73	37	97	E	109
5	F	17	33	65	33	97	49	129	F	145
6	G	21	41	81	41	121	61	161	G	181
7	H	25	49	97	49	145	73	193	H	217
8	I	29	57	113	57	169	85	225	I	253
9	K	33	65	129	65	193	97	257	K	289
10	L	37	73	145	73	217	109	289	L	325
11	M	41	81	161	81	241	121	321	M	361
12	N	45	89	177	89	265	133	353	N	397
13	O	49	97	193	97	289	145	385	O	433
14	P	53	105	209	105	313	157	417	P	469
15	Q	57	113	225	113	337	169	449	Q	505
16	R	61	121	241	121	361	181	481	R	541
17	S	65	129	257	129	385	193	513	S	577
18	T	69	137	273	137	409	205	545	T	613
19	U	73	145	289	145	433	217	577	U	649
20	X	77	153	305	153	457	229	609	X	685
21	Y	81	161	321	161	481	241	641	Y	721
22	Z	85	169	337	169	505	253	673	Z	757
23	2 A	89	177	353	177	529	265	705	24mo.	
24	B	93	185	369	185	553	277	737	B	1
25	C	97	193	385	193	577	289	769	C	49
26	D	101	201	401	201	601	301	801	D	97
27	E	105	209	417	209	625	313	833	E	145
28	F	109	217	433	217	649	325	865	F	193
29	G	113	225	449	225	673	337	897	G	241
30	H	117	233	465	233	697	349	929	H	289
31	I	121	241	481	241	721	361	961	I	337
32	K	125	249	497	249	745	373	993	K	385
33	L	129	257	513	257	769	385	1025	L	433
34	M	133	265	529	265	793	397	1057	M	481
35	N	137	273	545	273	817	409	1089	N	529
36	O	141	281	561	281	841	421	1121	O	577

American Point System.—Messrs. Mackellar and Co. publish the following on the 12-point = pica system:—

Point Body.		Old Name.
3	=	Excelsior.
$3\frac{1}{2}$	=	Brilliant.
4	=	Semi-Brevier.
$4\frac{1}{2}$	=	Diamond.
5	=	Pearl.
$5\frac{1}{2}$	=	Agate.
6	=	Nonpareil.
7	=	Minion.
8	=	Brevier.
9	=	Bourgeois.
10	=	Long Primer.
11	=	Small Pica.
12	=	Pica.
14	=	{ 2-line Minion. / English.
16	=	2-line Brevier.
18	=	{ Great Primer. / 3-line Nonpareil.

Point Body.		Old Name.
20	=	{ 2-line Long Primer. / Paragon.
22	=	2-line Small Pica.
24	=	2-line Pica.
28	=	2-line English.
30	=	5-line Nonpareil.
32	=	{ 3-line Small Pica. / 4-line Brevier.
36	=	{ 2-line Great Primer. / 3-line Pica.
40	=	Double Paragon.
42	=	7-line Nonpareil.
44	=	{ 4-line Small Pica. / Canon.
48	=	4-line Pica.
54	=	{ 5-line Small Pica. / 9-line Nonpareil.
60	=	5-line Pica.
72	=	6-line Pica.

Margins for Books.—The tendency of the earlier part of the century, by which we were given liberal margins to books, now seems to be much altered, and the width of the pages has been materially diminished. It would seem that publishers are now anxious to get as much upon the leaf as possible, and every expedient is resorted to for that purpose. This is not only the case with publications like magazines, or with heavy volumes like cyclopædias, but has become the rule with histories, biographies, and critical works. It deprives the binder of the paper necessary for him to use to make the appearance of the book symmetrical, and forces him to place the reading matter so near the back of the leaf that it is with difficulty the book can be opened wide enough to allow it to be read. This is not as it should be. No book should be imposed or worked without allowing sufficient space between the pages for all the exigencies of the binder, and, except in very small or thin books, the half-inch or so which is allowed is insufficient. The calculation for space must be made when the plan is first laid out, and no considerations of economy should be permitted to interfere with a liberal allowance. The proportion which good books ought to have was settled by the early French, Italian, and Dutch printers two centuries ago, and has since been always followed by those who know anything of their business. The top and inner margins are comparatively smaller than those at the outside and bottom of the page, which are very liberal. In this way the shears of the binder occasion no great destruction, and to this action of the celebrated typographers and binders of early days we owe the preservation of the works upon which they bestowed so much care, which by our modern careless usage would have tumbled into pieces of themselves in a few years.

The Distribution of Jobbing Types.—The plan adopted by many large printing firms, of having one person attend to all the job-type distribution of an office, is a sensible one, and we wonder that the rule is not more generally observed. It would doubtless be a difficult matter to find two establishments that lay their cases alike; and this is one of the important reasons why it does not pay to intrust the distribution of valuable job-type to every hand that may be put to work. The misplacing of a few characters of a fount would, in some instances, render it entirely useless. This is not an uncommon occurrence in large printing offices. A printer will often spend more time hunting through the cases for missing letters than it takes him to set the job. Display sorts not only cost much, but it takes time to procure them.

Cutting Reglet and Furniture.—Never cut up short pieces to make shorter ones. This leads to great waste. But when a piece of the required length cannot be found, cut one off a new length and put it away carefully in its order. If this plan is followed, in a short time there will be plenty of pieces of all lengths without resorting to the saw, while on the other plan you are continually cutting and never have an adequate supply.

A Place for Cuts.—The too common practice in printing offices of dropping cuts and stereos. of all kinds "wherever it comes handy,"—on top of racks, in sort cases, or under the stone,—is not only slovenly, but it is inconvenient and wasteful. The time lost in hunting them up under such circumstances, in a moderately busy office, would soon pay for a cabinet for their reception. In every well-regulated printing office there should be a suitable receptacle for these blocks, properly labelled, and if catalogued and indexed, so much the better. If the cuts aggregate a large number, a cabinet, especially made if need be, should be provided, where every one should have its appropriate place, to be kept there when not in use.

A Plea for New Founts.—It may be urged that the frequent buying of type loads an office down with too many styles; that cases cost money as well as type: and that, finally, the investment becomes greater than the business warrants. But we do not believe in retaining all the old or obsolete things, for they become unprofitable tenants after a time. Clear out all the oldest founts, making case room for the new. The old type has paid its way and served its purpose well; but now it is rarely, if ever, used, and its value as old type, towards the purchase of new, will be plainly greater than the idle occupancy of the cases. By rotating carefully in this way the cost will not be felt.

Blocks for Tinted Grounds.—These are made, as is well known, of various materials, such as brass, type-metal, box or other wood, celluloid, etc., and a printer in any great commercial centre is always able to satisfy his requirements almost at a moment's notice. Not so his brother in the provinces, who may be seriously inconvenienced when one of his customers takes a fancy to colours. The way small German printers help themselves may, therefore, be read with advantage. They take a very smooth and even glazed board, such as is used in pressing and glazing the paper after printing, and cut two or three pieces out of it, a little larger than the desired tinted block. A similar piece is cut out of common pasteboard, well sized and smooth, and the whole is then formed into a layer with best glue, and stuck, also with glue, on a solid wooden board—oak will suit best—so as to form a type-high block. This done, the block is put into a glazing press, or into a letter-copying press, or you may simply put boards and heavy weights on it, just to let it dry under severe pressure, care being taken that the pasteboards are well united to each other and to the wooden block, as the result depends on it. When the wood and paper block is completely dry, the transfer may be made. The forme to which the tinted block is to fit is locked up in a frame, as also is the block itself, the latter being disposed so as to fit in the composed matter as exactly as possible. Then this last one is lifted into the press or machine, and, after being well inked, one pull is made from it on to a sheet fixed on the tympan or cylinder. This done, the formes are changed, the block forme taking the place of the composition forme, after a sheet of thin pasteboard, of the thickness of common Bristol board, has been put on the bed of the press, to raise the block a little above type height to give more effect to the pull. The tympan, still bearing the impressed sheet, being now brought down and a pull taken, a negative copy of the contents of the forme will be obtained on the glazed-board block, the cutting of which may now be proceeded with. That operation is best effected with a thin pen-knife, or even with a small chisel, care being taken to cut in an outward slanting direction, to give the printing surface of the block a larger and stronger basis. When the cutting is finished, the block is ready for printing; but should the number of copies to be printed from it be large, it is better to give it first a coating of varnish. A coating of shellac, diluted in alcohol, applied twice, has proved most effective, and will stand the printing of 10,000 copies before any change is to be observed. In cleaning, lye must be avoided, and only a little petroleum or turpentine rubbed over with a smooth rag. When the tinted ground is to show a pattern, this may be obtained by sticking embossed paper on it, taking care to fill the indentations of the paper well with stiff glue, and to paste it thus on the block. When dry, it may be varnished and printed from as before stated.

Metal Tint Blocks.—Blocks of type-metal are often used in printing tinted grounds, but prove dangerous to fine and delicate-coloured hues, as dirt may be on the surface without being visible to the eye. To keep them perfectly clean is therefore a strict necessity, and that is best arrived at by sprinkling them with benzine and wiping it very smoothly off with a soft bit of rag. Hard rubbing entirely to be avoided, as it will immediately soil the metal-blocks.

Proper Names.—As an interesting and valuable comparison of the number of capital letters required for an average collection of proper names, the following table has been published of names in the " New York City Directory," containing 313,992 names beginning with each letter of the alphabet :—*A*, 7,643; *B*, 29,721 ; *C*, 21,808 ; *D*, 16,016 ; *E*, 5,971 ; *F*, 14,408 ; *G*, 15,560 ; *H*, 24,842 ; *I*, 1,106 ; *J*, 5,429 ; *K*, 15,798 ; *L*, 15,097 ; *M*, 34,048 ; *N*, 5,604 ; *O*, 5,924 ; *P*, 10,540 ; *Q*, 919 ; *R*, 15,927 ; *S*, 33,652 ; *T*, 8,098 ; *U*, 1,114 ; *V*, 3,728 ; *W*, 18,663 ; *X*, 10 ; *Y*, 913 ; *Z*, 1,453.

Electro-Matrice Type-Founding.—Electro-matrices, as used by some founders, are made in three parts of brass and copper combined, through the agency of an electro-battery, and are liable to swell or vary in line, height, and thickness; not having an angle to assist the type in its delivery from the matrice, is smaller than the original, has a rough-looking appearance on the sides and back nearest the face, and in every respect inferior to type produced through the agency of steel punches driven in copper.

An Elastic-Faced Printing-Type is a recent invention constructed upon an entirely new plan, which involves the least possible amount of wear and friction, prints with the greatest ease, and insures the most perfect results yet attained. It consists of a hard-bodied printing-type, whose printing character is made integral with the body. An elastic coating or cushion is moulded and vulcanized to the type body, the character projecting into the elastic coating, forming an elastic-faced printing character, which is supported and secured firmly in place by the type.

Abstract of the London Compositors' Book Scale.

					Dictionaries.					Catalogues.			Greek.	
			Common.	Foreign.	English.	English.	Foreign.	English Grammars.	Foreign Grammars.	Library.	Book-sellers'.	Auc-tioneers'.	Without Accents.	With Accents.
			d.	*d.*	*d.*	*d.*	*d.*	*d.*	*d.*	*d.*	*d.*	*d.*	*d.*	*d.*
Manuscript.	English to Brevier	*leaded*	6¾	7¾	7¾	8¼	8¾	7¾	8¼	7¼	8¼	7	9½	11
		solid	7½	8½	8½	9	9½	8½	9	8	9	7½	10¼	11¾
	Minion	*leaded*	7	8¼	8	8½	9	8	8½	7½	8½	7¼		
		solid	7¾	9	8¾	9¼	9¾	8¾	9¼	8¼	9¼	7¾		
	Nonpareil	*leaded*	7¾	9	8¾	9¼	9¾	8¾	9¼	8¼	9¼	8		
		solid	8½	9¾	9½	10	10½	9½	10	9	10	8½		
	Ruby	*leaded*	8¼	9½	9¼	9¾	10¼	9¼	9¾	8¾	9¾	8½		
		solid	9	10¼	10	10½	11	10	10½	9½	10½	9		
	Pearl.........	*leaded*	8¾	10	9¾	10¼	10¾	9¾	10¼	9¼	10¼	9		
		solid	9½	10¾	10½	11	11½	10½	11	10	11	9½		
	Diamond ...	*leaded*	10¾	12	11¾	12¼	12¾	11¾	12¼	11¼	12¼	11		
		solid	11½	12¾	12½	13	13½	12½	13	12	13	11½		
Reprint.	English to Brevier	*leaded*	6	7	7	7½	8	7	7½	6½	7½	6¼	8¾	10¼
		solid	6¾	7¾	7¾	8¼	8¾	7¾	8¼	7¼	8¼	6¾	9½	11
	Minion	*leaded*	6¼	7½	7¼	7¾	8¼	7¼	7¾	6¾	7¾	6½		
		solid	7	8¼	8	8½	9	8	8½	7½	8½	7		
	Nonpareil	*leaded*	7	8¼	8	8½	9	8	8½	7½	8½	7¼		
		solid	7¾	9	8¾	9¼	9¾	8¾	9¼	8¼	9¼	7¾		
	Ruby	*leaded*	7½	8¾	8½	9	9½	8½	9	8	9	7¾		
		solid	8¼	9½	9¼	9¾	10¼	9¼	9¾	8¾	9¾	8¼		
	Pearl.........	*leaded*	8	9¼	9	9½	10	9	9½	8½	9½	8¼		
		solid	8¾	10	9¾	10¼	10¾	9¾	10¼	9¼	10¼	8¾		
	Diamond ...	*leaded*	10	11¼	11	11½	12	11	11½	10½	11½	10¼		
		solid	10¾	12	11¾	12¼	12¾	11¾	12¼	11¼	12¼	10¾		

Reprints not in every respect exact reproductions of the originals, are cast up ¼*d.* per 1000 extra; Reprints with MS. insertions ½*d* per 1000 extra.

Stereotyped matter with high spaces is cast up ¼*d.* per 1000 extra; Stereotyped matter with low spaces is cast up ½*d.* per 1000 extra.

Thin founts are cast up ¼*d.* per 1000 extra for every en below 12 ems of their own body.

Bastard founts of one remove are cast up to the depth and width of the two founts.

Locking up Formes for the Foundry.—Great annoyance and danger is often experienced in electrotyping from formes which have not been properly locked up. A proof taken from a forme after the quoins have been pushed up with the fingers may show every line straight and everything in its proper place, when, after finishing the lock-up, the tightening of the quoins with the mallet and shooting-stick or wrench will produce displacement of lines and types alike provoking and unsafe. Formes for electrotyping should be locked up even more tightly than when sent to press ; for the adhesiveness of the wax mould is more likely to draw letters than the suction of rollers. The forme should be well planed down after the final locking up—care being taken that no dirt or other extraneous substance is under it—to insure its being perfectly flat. A proof of every forme should be taken after the final lock-up, and this proof carefully examined to see that nothing has been dropped or displaced. A clean proof should be sent with the forme to the foundry, that the electrotyper may have an opportunity to examine it, and to repair any damage that may possibly happen to the forme while in his hands. Before locking up a forme, type-high bearers at least a pica in width should be placed around it, and between (and close to) the several pages, if there are more than one. Open spaces, as about headings, blank pages, etc., should also have bearers made by placing types in blank spaces. These bearers serve the double purpose of guards for the matter in the formes, and enable the workman to have something to rest the plates upon when finishing them.

Ordering Sorts.—The twelve square boxes directly in front of the compositor, containing the letters a, c, d, i, m, etc., will hold about 2 pounds each. The boxes half the size of the "a" box will hold 15 ounces each, containing the letters f, b, g, l, p, etc. The small square boxes containing the letters k, j, q, etc., will hold 6 ounces each. The "e" box 3 pounds, and the cap case 5 ounces to the box. The best way to order sorts for display type is to do so by "irons." A typefounder's "iron" is about 20 pica ems long.

Title-Pages.—A writer contrasts the title-pages of to-day with those of the past, to the decided disadvantage of the former. There can be no doubt that there was much more character and individuality in the old than in the new. The disposition of the type was less conventional; and the illustrative borders and the like had a picturesqueness of which the glory is now altogether gone. Title-pages certainly used to have a quaintness, and usually a special appropriateness, to which they can now lay no claim. They had a meaning of some sort; they were the elaborate portals through which the reader passed to the books themselves, foreshadowing in some respects the matter which there awaited him. The modern title-page is a very much simpler affair, and for that reason alone, may commend itself to the majority. It is at least clearly set forth and readily mastered—save, perhaps, in a few exceptional instances, which do not count. *Simplex munditiis* might be the well-deserved motto of the average title-page as we now see it. Nor, for the greater number of volumes, would one particularly care to welcome any more elaborate treatment. In the old times books were few and precious; they were valuable, and were valued. Nowadays only a minority are *éditions de luxe*. The bulk are produced less for beauty and permanence than for utility and the moment. On such it would be waste of money and care to bestow title-pages other than the plainest and the neatest. Books are now multitudinous, and we are a business people. What, therefore, is of the first importance is that the title-page of a volume shall contain little print, and be easily conned. A plea may be put in, on behalf of the bibliophile, in favour of the invariable presence on the title-page of the date of publication, for, without that, a book is like a child without a record of its birth or baptism.

Cork Tint Blocks.—This makes an excellent tint, cut in thin strips and mounted on block, worked over light background, and bronzed with gold, copper, maroon or fire bronze. The small punctures which are found in cork, when printed in this way, produce a very odd and unique appearance. Very pretty for cards, folders, or run across corners, etc.

How to Cast-off Copy.—Although entirely exact rules for casting-off copy cannot be laid down, the following may be recommended as the result of experience. After having made up a composing stick to the measure proposed for the width of the work, to take an average page of the copy, and set from it until a certain number of lines of the manuscript come out even with a number of lines of types. From this a calculation can easily be made for the whole of the work. Suppose a manuscript of 250 pages, and 31 lines in a page, be brought into an office, and it is required to determine how many pages it will make in long primer, the page being 28 ems wide and 40 lines of types in length; and it is found, by setting up a few lines, that 9 of the manuscript are equal to 7 of the types. Then:—

$$\begin{array}{l} 250 \text{ pages manuscript.} \\ 31 \text{ lines in a page.} \\ \hline \\ 250 \\ 750 \\ \hline \\ 7750 \text{ lines manuscript.} \end{array}$$

$$9 : 7750 : : 7$$
$$7$$
$$\overline{}$$
$$9 \,) \, 54250$$
$$\overline{}$$
$$40) \, 602,7 \text{ lines of types.}$$
$$\overline{}$$
$$151 \text{ pages of types.}$$

The number of sheets can be ascertained by dividing 150 by 8, 16, or 24, according to the size of the signature in which the work is to be printed.

Origin of Roman Type.—This character with lower-case, modelled after the cursive writing of the twelfth century, was first reduced to symmetry and used as a body type for book-work in 1471, by Nicholas Jensen, a famous printer of Venice.

Special Sorts.—Where do they all go to—such as fractions, reference marks, accents, italics, etc. ? Every employing printer who has had much book or news work that called for any quantity of these sorts, has had the problem put sharply to him by the frequent demand for fresh supplies. On inquiry, he has traced their disappearance to the laziness —it is nothing more nor less—of piece-hands, who, not thinking that these sorts are likely in an emergency to be very valuable, either throw them into the quad-box, or scatter them promiscuously and recklessly about in the upper case.

Lock-up for Galleys.—A cheap, effective end-lock is a piece of flexible steel, lead high, bent between side of galley and side-stick. This can be used quickly, and is effective on any size galley.

Fashions in Type.—There are fashions in type as in modes of apparel; and it is unwise to attempt to stem the tide of fashion in these as it is foolish affectation for one to wear knee-breeches and powdered wigs to-day. But it is unnecessary to follow fashion to the extreme ; and the small establishments can no more keep up with all the various styles than a printer's wife can dress like a millionaire's. There are certain standard faces of type that cannot be dispensed with in any office ; but it is not advisable to choose the oldest cuts of these. One founder may have the matrices for the earliest styles of full-face, antique, etc. ; but many of the latter shapes are vastly to be preferred as being clearer cut and better proportioned. Punch-cutting has greatly improved of late years, and it cannot be denied that a more artistic taste prevails than existed half a century ago. Even our plain romans are much improved, as will be seen on a comparison between the newspapers and books of 1834 and those of a later date. Some of the " improvements " in types, however, have been in a backward direction, as witness the grotesque and almost illegible styles that have appeared of late. It is not necessary that these ugly shapes should have place, and many of them should be utterly tabooed.

The Parts of a Type Named.—The *face* is the part from which the impression is taken. The *counter* is the sunk space existing between the lines of the face. The *nick* is an indentation cast in the body of the type. This is always placed on the side leading up to the bottom of the face. The *shoulder* is that portion of the type not occupied by the face. The *beard* comprises the bevelled bases running from the shoulder to the face. The *pin mark* is a small circular indentation on the side of the body. This is formed by the pin which holds the type in the upper half of the mould, and acts as a drag to deliver the type. The *height to paper* is the distance from the face to the feet of the type. The usual height is eleven-twelfths of an inch. The *body* of a type is the thickness from back to nick. The *feet* are the portions supporting the type, each side of the groove, and are formed in the process of planing off the superfluous metal left at the point where the jet has been removed.

Type Cases.—Printers' cases should always be selected with a view to accommodating their type in the most accessible manner, and without crowding. Delicate faces, like scripts, should never be laid in cap cases, or crowded into the boxes, nor should founts without lower case be laid in italic cases. Never lay two founts of type in the same boxes. The time wasted in setting it out is soon enough to pay for another case. Use cap or triple cases for all-cap founts, according to their size, and do not under any circumstances lay delicate faces with heavy type. Never crowd type together. It is not only disastrous to the faces, but is a loss of time in setting. Every printing office should have, as a part of its regular quota of cases, a figure–case for extra figures, a space and quad case containing all sizes for spacing job work and advertisements, and blank cases for cuts, etc. These are no luxuries, but the best of investments, and any printer who purchases them will find that he is amply repaid in a short time. "A place for everything, and everything in its place," applies with greater force to a printing office than any other line of business in the world. It is made up of numberless articles and appliances, any one of which is liable to be called into use at a moment's notice. The most perishable and costly portion, with the exception of fine type, are the cuts and electros. They should, therefore, be given a safe and convenient place of storage. The blank cases, which will fit into any frame, are the cheapest for this purpose.

A Suggestion for Type-Founders.—A great source of worry to the painstaking printer is the ugly gap that so often occurs in words in lines of capitals where the letters, A, W, Y, etc., come into juxtaposition, particularly in large round-faced founts. A great deal of time is often spent in dove-tailing such letters by means of rude implements—such as a saw, an old file, or a pocket knife—generally available in the average job office. The suggestion is made to progressive type-founders that a few in each fount of the letters (caps) A, F, L, P, T, V, W, Y, be cast or cut with a mortice in the open portion of them, so that they fit into one another when coming together in a word. The word RAILWAY, for example, would be much improved in appearance if the last four letters came closer together by means of such mortising.

Roman Type in Germany.—Roman type and script are making their way slowly but surely in Germany. The society for the abolition of the old German letters, which in 1866 numbered only 2,871 members, now has 4,436 on its list, which includes teachers, physicians, booksellers, and merchants. In the last year thirty-one professors joined the league—a notable fact. In 1886, according to Heinrich's "Bibliography," out of 6,213 books on artistic, scientific, and mercantile subjects 5,316 were printed in roman type. As is well known, Bismarck has set his face stubbornly against this reformation, which will, when fully accomplished, be for Germany, as well as for the rest of the world, a real blessing, inasmuch as their barbarous system of type and script has, as it were, shut the world out of their domain of letters.

Scarce Hair Spaces.—If you find yourself short of these, a run round the display founts will generally produce the desired result. They are often dropped with the letters in distributing.

Type made from Paper.—Type made from paper is the latest novelty. A process has been patented by which large type used for printing placards can be made from pulp. Such letters are at present cut on wood. The pulp is desiccated and reduced to a powdered or comminuted state, after which it is thoroughly mixed with a waterproofing liquid or material—such as paraffin oil or a drying linseed oil, for instance. The mixture is then dried, and subsequently pulverized. In its pulverized state it is introduced into a mould of the requisite construction to produce the desired article, type, or block, and then subjected to pressure to consolidate it, and heat to render tacky or adhesive the waterproofing material. Finally, the type is cooled while in the mould, so as to cause it to retain its shape and solidity.

A Mount for Illustrations.—In a recent issue of the "Superior Printer" an imitation passepartout frame or mat is printed round a picture. The editor thus describes how it is done. He first took an old electrotype block and pasted a piece of very coarse sand-paper upon it. After it was entirely dry he took a ruler and sharp penknife and trimmed the sand-paper to the size desired. Then he locked the block in a chase, and after taking the rollers out of the press and preparing the tympan he took a brush full of paste and daubed the tympan thoroughly where the block would come in contact with it. Then he laid over the tympan a sheet of soft book paper, then daubed it with paste again and laid on top of that another piece of soft paper. Then he took an impression upon the tympan and allowed the press to remain so for about half an hour, when the guides were set and the job run off very quickly. Cardboard can be treated in the same manner. Some very good effects can be obtained by using plates made from book-cloth in place of sand-paper. The reader will find that it is a great advantage to be able to make one's own "Rufenuf" (rough-enough) papers and cards, as you can do your printing first and then emboss your job afterwards, thus avoiding the trouble of trying to do a decent job of presswork upon the rough-surfaced papers and cards furnished by the paper-dealer.

Patent-Leather Tint Blocks are said to produce most excellent results in letterpress printing. It must be explained, however, that patent-leather is meant and not leather that is somebody's patent. Patent-leather, a sort of canvas with a varnished leather face to it, is over here known as American cloth. Use the best and thickest that can be obtained and glue a piece rather larger than the size of the tint block required to a wooden background: squeeze out air bubbles and keep under pressure until set. Trim down the leather to the required size, and the tint block, which is almost everlasting and gives results beautifully clear and uniform, is complete.

Justification of Curved Lines.—When it is necessary to put a curved line in a job, set the curved line first, and secure it so it will lift and hold together independently of outside pressure. This can be done by setting it in a patent brass curve and clamp, but when these are not available, set the line in curved leads. When the line is accurately spaced, carefully read to see that it is all right, and each letter curves properly, take a piece of paper, put mucilage on it, and apply the mucilage to the line. Put a little mucilage between the paper and the curved lead. When this dries your line will lift like a slug. Before distributing such lines, let them soak in water, kept in a saucer, and then carefully clean off the mucilage. This practice will save lots of time when the forme goes to press, and if properly done will insure a perfect curve.

New Margin Gauge.—A M. Anthinous, of Caen, has invented a new gauge for margin-making, to be used by clickers and others intrusted with the making-up of formes of book-work. His " Margeometer," as he calls it, is a sheet of stout paper on which is lithographed a large triangle. This triangle is divided by transverse lines ruled at equal distances, and these are crossed by other parallel lines, figures being inserted at the points of intersection to facilitate calculation. " To determine the furniture necessary for any forme, the printer folds a sheet of the paper to be used for the work according to the forme, and places it on the Margeometer, the back to the left hand and the head upwards. He now marks

E

the place of the first line to the left which bears a number, and thus ascertains the width of the inner margin; then he marks on the right hand the third line bearing the same number which shows him the width of the outer margin." The inventor's own description of the mode of using the apparatus, appears to us not very lucid; but the Margeometer must be meritorious, as it has not only been praised by the French printers' papers, but has won for its inventor a silver medal.

Recipe for Owltype.—Take a very heavy coloured railroad board, or thick glazed mill-board, and if very smooth pass a damp sponge or rag over the surface, to take off any heavy gloss. Then take China white, or any fine clay, mix it with water to make a paste about as thick as common molasses, adding a few drops of mucilage to a wine-glass full of the softened clay. The less mucilage the better, though some is required to prevent the clay peeling off the cardboard too easily. Use a rag, spatula, or small pencil brush in applying the softened clay to the cardboard, of course leaving spaces or spots on the cardboard uncovered by the clay. These spots will produce the solid parts when printed. The clay dries rapidly, so that the matrix may be placed in the stereotyping backing-up pan and cast almost immediately. Then block and trim and put on press.

"Chaostype" and "Selenotype" is said to be quite simple, and consists of pouring quickly, but not continuously, melted metal into a *cold* stereo casting-box in which the shape required has previously been arranged, with the help of core bars, or the whole box can be filled and cut to sizes afterwards. We are informed that is the sole mystery, but the various "chaotic" patterns are the result of practice in dropping in the metal.

Hint for Case Racks.—Leave an empty place in all long case racks, at a convenient height for placing a case out of such rack in which to distribute a line or two, or to set up a line, instead of having to carry such case away.

Relative Sizes of English and German Type Bodies.

—The following table, although not strictly accurate, will be found useful.

BODIES.	Equivalent in 8 to pica leads.	Equivalent in Didot points.
4-line pica	32·00	45·00
Grobe Canon (German)	29·90	42·00
Doppel Text (German)	28·48	40·00
2-line double pica	27·75	39·00
Kleine Canon (German)	25·62	36·00
3-line pica	24·00	33·75
2-line grt. pmr. Doppel Tertia (Ger.)	22·75	32·00
Doppel Mittel (German)	19·92	28·00
2-line English	17·95	25·00
Doppel Cicero (German)	17·08	24·00
2-line pica	16·00	22·50
Text (German)	14·24	20·00
Double pica	13·87	19·50
2-line long primer Doppel bourgeois (Ger.)	12·81	18·00
Great primer. Tertia (German)	11·37	16·00
2-line brevier	10·32	14·50
Mittel (German)	9·96	14·00
2-line minion	9·25	13·00
English	8·97	12·50
English [cicero]. Cicero (German)	8·54	12·00
English [small]	8·42	11·84
Pica	8·00	11·25
Pica (American)	7·92	11·19
Brevier (German)	7·83	11·00
Corpus or Garmond (German)	7·12	10·00
Small pica	6·99	9·75
Long primer. Bourgeois (German)	6·40	9·00
Bourgeois. Petit (German)	5·69	8·00
Brevier	5·06	7·25
Colonel (German)	4·98	7·00
Minion	4·63	6·50
Emerald	4·48	6·25
Emerald [nonpareil]. Minionette (American). Nonpareil (German)	4·27	6·00
Emerald [small]	4·21	5·92
Nonpareil	4·00	5·62
Perl (German)	3·56	5·00
Ruby	3·49	4·82
Pearl	3·20	4·50
Diamant (German)	2·84	4·00
4 to pica	2·00	2·81
6 to pica	1·33	1·87
8 to pica	1·00	1·40

Weight of Type for Newspapers.—Country printers often ask how to estimate the quantity of type necessary for a paper of given dimensions. The following will be found a correct and simple plan :—A page of type 4 in. by 6 in. weighs $7\frac{1}{2}$ lbs. Let this be taken for a starting point, and the weight of type when set up and ready for imposition will easily be found. Then add 40 per cent. to the weight he arrives at to cover inequalities of sorts, and the letter necessarily lying in case, and this will be found near enough for all practical purposes. At the same time, make an allowance for over-matter and possible supplements.

Border Cases.—Keep these " up," on frames handy to every compositor, and not. as is frequently the case, on some one compositor's frame—both in his own way and not at all get-at-able by the other members of the staff.

Hints on Composition.—Understand your take fully before leaving the foreman or copy hook. Time spent in this way is profitably invested. At least read through the outlines of the job. If pamphlet or book-work, the reading of the first page or two will be sufficient. Formulate your plan of development. Determine upon display lines. Spelling, style of punctuation, capitalizing, and paragraphs, should be according to usage of establishment. If possible, absorb the subject of your take ; it will render work more engaging. As to rapid composition, absolute oblivion to surroundings is essential. Like an actor or orator, you should mentally get inside of the subject ; shut your own other senses, and utilize that only which is necessary to rapidity and correctness. Some have a new sense created by rapid composition, combining mental and physical phenomena, rare and wonderful. As in distributing, stand square on your feet, with chest distended. Hold stick well in front, so as to be in full view of left eye, while the right generally is manipulating movement of picking up letter and reading copy. Type should be grasped with the right-hand thumb and forefinger, with a sliding approach, so as to lift with finger and balance with thumb. After catching the word with the eye and mind, concentrate on the immediate letter to be picked up, with an active plunge of the hand toward the box, without the pressure of nerve force if possible ; bring back letter swiftly to stick, striking rule as near as possible to location of word. Seize letter with left thumb and strike out with right hand immediately for next letter. The casting of hand into box, seizing letter without hesitancy, and the withdrawal to stick, should be of same velocity. The movement, physical and mental, generally determines the speed of the compositor. Rapid composition comes from mental anticipation coupled with will power. It can be cultivated.

Easy Method of Cutting Wood Letters.—A continental printer is manufacturing wood type by a very simple process. He sticks the printed type he wishes to cut out on a thin slice of wood, or draws it on with a pencil, using a corksaw to cut out the shape of the type, even out of several woodslice at once, putting one upon the other, afterwards fastening the type on a solid block of wood corresponding with its size. The finishing touch, to do away with any roughness, is given with the graver, penknife, or rasp.

Distribution Galleys.—Do not distribute matter from mahogany galleys, but lift on metal galleys before wetting the matter. Galleys are soon ruined by carelessness in regard to this. At the same time do not damp the type on the frame.

Sectional Type was first introduced into this country in 1878 by Messrs. Caslon and Co. It consists of a condensed sanserif, each letter being cut into two parts across the middle, which allows of the introduction of another line. It is quite distinct from mortised types, which are not cast to the ordinary rectangular shape, but are cut away in some of the open parts to allow of other letters being brought up close. Mortised types were invented by Mackellar, Smiths, and Jordan, of Philadelphia, and patented in 1884.

A **New Rule-Cutter** has been invented to replace the ordinary rule-cutting machine. It is adapted for cutting brass and metal rules of all thicknesses, as well as reglets, side-sticks, etc., and can also be used for trimming stereotype plates. The article to be cut is firmly held in position by the aid of a strong lever, while a saw working in a groove. and so mounted as to always cut at a true right angle, performs the cutting operation. When in use the teeth of the saw require repeated oiling with a brush, and the invention is said to give great satisfaction with the expenditure of little effort. The cut is claimed to be sharp and clean, and in the case of brass rules not to need further finishing.

MACHINE AND PRESS WORK.

PROCESS Blocks.—The question is frequently asked, "How many impressions can be printed from half-tone blocks?" We think it will interest many of our readers to learn that 135,000 good copies have been printed from the Meisenbach blocks supplied for the Christmas number of the "Lady's Pictorial," still leaving the blocks in a perfectly satisfactory condition.

A Useful Wrinkle.—A printer complained that he could not make a colour job register on a Universal press. One or two questions elicited the information that the gripper movement had become so badly worn that the grippers " crawled " after striking the platen and threw the sheet out of register. Many printers have encountered the same difficulty, and will be interested to know what remedy was prescribed, which was simply this : Take two pieces of strong cardboard, each about two inches long, bend them near the centre, and with mucilage or paste secure each to the tympan sheet, so that the tongue, or the free end of the card, overhangs the sheet at a point beneath the gripper. These tongues, striking the sheet first, and being fixed to the tympan, will prevent any movement of the paper by the grippers, and insure accurate register. This is a simple expedient, and costs nothing, but stow it away in one of the cells of your memory—it may be useful some day.

Cylinder Packing.—Never use rubber or other soft packing, except for old type. Hard packing for new type should be the rule, and do not pack deeper than will be level with the bearings of the cylinder. Use a couple of sheets of book paper for the outside.

Gluing Machine Belting.—As a rule, belting for machines is still sewn together in most printing offices. Abroad, the gluing of the belts has been found to answer very well. The double belts are split up a few inches, say about three, at both ends, the upper part of the one, and the under part of the other cut away, and both overlying parts glued together, which can be done easily in a few minutes. When the belt is drawn through the machine, the two ends of the belt are seized, the belt is brought to the required degree of tension and then glued. Single thin belts for steam-presses can thus be glued instead of sewn.

Rapidity in Making-Ready.—An American journal says:—What is the secret of speed in making-ready? Pressmen who have not become skilful at it wonder why others get their formes " up " quicker than they; and being unblessed with observant eyes or reflective minds, they jog along in their old ruts, while their fellows pass them in the race. There are three kinds of slow pressmen: (1) those born with thick wits, (2) those with lazy muscles, (3) those with active hands and brains, but who have had no chance to see how things are done by the rapid workers of the craft. For the benefit of the third class, we will point out the secret of success in obtaining a rapid make-ready. This secret lies in the study of the chief inequalities of each forme, and in rectifying those by underlays before a moment's thought is given to the overlays or cut-outs on the tympan sheet. This is so simple that the inquiring reader may pooh-pooh it as a paltry key to unlock so big a door ; but we assure him that it is the one key which, if he will use, will let him out of the dark hole he is in. How often have we seen a pressman fussing and fuming with an overlay—pasting here and cutting there hour after hour—sweatingly conscious that the foreman or employer judged he was consuming too much time ! Had he first levelled up his forme from beneath, he would have been surprised to find how little was left to be done on top. He would have discovered that he had brought up everything where it could be touched and thoroughly inked by the rollers. The neglect of this precaution is the cause of nine-tenths of the trouble from stoppages for patch-work, in addition to the original waste of time from a false system of making-ready. The overlay is for finishing, precisely as the cut-out is. Both should be employed only for delicate differences of impression, never for serious ones. To the second class in our category—the lazy ones, we would say: Guard against the indolence that would prevent you lifting the forme, for the fault will not only cause you a waste of time, but will infect everything you do. It is the active, willing workman who gets along. Never shirk! It doesn't pay.

Working of Blue and Green.—If you want to get the best results with blue ink, especially ultramarine, as well as the more brilliant greens, don't use hard rollers, nor glycerine rollers. Use good fresh glue-and-molasses rollers, and do not carry too much colour. It is common to notice a mottled or speckled appearance in solid blue surfaces. In other cases a stringy appearance is seen. This indicates either too much moisture or too much oil in the rollers—a simple matter, but one to be watched with care.

Dividing the Ductor for Colour Work.—To make a good, cheap and serviceable fountain division for working two colours at once: Take a piece of hard soap and cut as near the shape and size of fountain as possible, push down to the fountain roller and friction will soon fashion the soap to hug the roller so nicely that all danger of the mixing of colours will be overcome.

For Bronzing on Plated Papers.—Pressmen frequently find that they cannot print plated papers in bronze with success. The heavy coating on the paper absorbs the size so that the bronze will not " stick." This can be obviated by running the sheets twice through the press, using size both times, and allowing it to dry after the first impression, which it will do very quickly. The first printing fills up the pores in the paper, leaving an excellent ground for the second impression, to which the bronze will adhere firmly. Of course this is expensive; but it must be figured in estimating the cost of the work.

Hints for Buyers of Machinery.—Messrs. R. Hoe and Co. give the following practical advice to intending purchasers of machinery: "In buying a machine, see that, whether new or second-hand, it is strong and well-made. Consider the standing of the maker, both as mechanician and machinist. A light-framed or shakily fitted machine will be dear at any price. Do not be deceived by any beauty of paint or finish on exposed work, which adds nothing to the usefulness of the machine, and which may draw the eye from an examination of the working parts. Uncover the boxes, and see whether the finish of shafts in their bearings, or journals, is as smooth and true as the white and brass work of more exposed pieces. Take out, here and there, screws and bolts: see if the threads are deep, sharp, and well-fitted. Look closely at the fitting of all toothed or pinion wheels; note whether they have been cast and filed to fit, or whether they have been accurately cut by automatic machinery, so that they will fit in any position. Slowly turn pinion wheels, and note whether there is any rattling or lost motion, or whether the teeth fit snugly, yet freely, so as to give even, steady motion. . Closely examine all castings for pinholes or air-bubbles, which may be most easily detected in work that has been planed. See that castings are heavy as well as solid. Look after oil holes and provisions for oiling. See that the castings are neatly fitted; that they do not show the marks of the hammer or file, which must be used to connect them if they have been forced or badly put together. Pay attention to the noise made by the machine when in motion: if fairly fitted, the noise will be uniform; if badly fitted, it will be variable or grating."

Printing on Tin-foil.—Put gum arabic in vinegar and let it stand until it becomes a heavy paste, then mix in with ink as varnish.

Bronzing after a Colour has been Printed.—Calcined magnesia rubbed on a job will allow of bronze being printed over colour without adhering to it, but the colour should be as dry as possible before applying the magnesia.

Difficult Distribution of Ink.—Proprietors of printing-offices would do well to inquire how the pressmen heat their slabs on a cold or damp morning. The common way is to roll a piece of paper—a clean sheet as a rule—light it, and then burn merrily under the slab. A more inflammable place than the press-room could not be imagined, for everything is of a combustible nature. It pays to have a few detachable gas tubes laid on, for it is nothing but gross carelessness to let such a practice continue.

Where shall Red Ink be Used ?—In ornate typography red is growing in favour, and the tendency is to work in heavy masses of it. To produce a striking effect more red is required than black. A recent number of the " Art Age," in an elaborate review of the use of red ink, says, among other pertinent things, that the mistake most frequently made is in introducing red inappropriately in masses where it is neither ornamental nor part of the general composition. To put it plainer, there is an increasing disposition on the part of printers who have a laudable desire to be progressive to use great masses of red merely for the sake of obtaining a glaring effect. In a circular of a recent sale of paintings this tendency was strikingly shown. A pretty red initial would have set off every page handsomely, but to the one ornate crimson letter was added an ugly, blotchy, fiery red head and tail-piece that, instead of rendering the red attractive, made it positively repulsive to the eye. No compositor of any taste or judgment would have overloaded a page with an ornamental initial and a head and tail-piece, nor would the man who designed the circular in question have ventured upon such an overweighted design but for the simple sake of introducing plenty of red. A single line of red in a page of Gothic produces a highly attractive effect. One heavy letter or line of red in a page is pleasing to the eye ; any further addition of red in mass becomes a positive blemish which repels.

To Protect Belting.—The anointing of leather belting with castor oil will prevent its being gnawed by rats.

The Damage to Rollers by Brass Rule.—In working brass rule formes, to prevent as much as possible the rollers from being cut, place broad wood rules inside the chases, so as to act as bearers for the rollers the whole length of the forme. The ends should be rounded or tapered, in order to prevent damage to the rollers when rising on to them. They should be longer than the formes, and must be placed exactly parallel with each other. For open or light matter this will be found an excellent plan, as the inking will be more even than when rollers are left to bump over the matter. Strips of paper must be fastened to the cylinder or platen for these rules to work on, and should be changed as required. Wood rules should be kept specially for this class of work.

Black Ink on Coloured Paper.—When using black ink on a tinted ground, or on coloured paper, it is necessary to observe, that the black changes colour in many instances, or loses its intensity. Printed on a blue ground, its strength and power are lost; on red, it appears dark green ; on orange, it takes a slightly blue hue ; on yellow, it turns to violet; on violet, it has a green-yellow shade ; and on green, it appears as a reddish grey. Printers should take heed of these peculiarities of black, or they may find their work worthless when done.

Working Cardboard.—Someone writes :—In working cardboard on a small drum cylinder, I am troubled by a slur on the last line, caused by the stiffness of the board, which prevents it from conforming readily to the curve of the cylinder, so that, as the impression ceases, the sheet flies out flat, making the job look dirty on the edge. I have obviated this by passing cords around the cylinder, fastening one end to the rod which holds the paper bands, and the other to be braced against, which the fly strikes, the sheet moving under the cords while being printed; but is there no better way? *Answer.*—Take one or more pins, according to the size of the job, cut them off, so as to make them type high, or a fraction over, which place in the furniture, so as to catch the end of the cardboard, and the slurring referred to will be prevented.

How Blotters are Printed.

—Printing on blotting paper presents several difficulties. The trouble which lithographers find in the printing lies in the fact that the absorbent paper does not lift the ink, the impressions appearing grey, broken, and at the same time filled up. The steam-press upon which the printing is done gets full of paper-dust to such an extent that the ink upon the rollers becomes pitch-like and the press itself is injured materially, as the fine dust enters all oil-holes and the machine becoming dry is subjected to unusual wear. Taking this into account a few hints respecting such work may come handy. It is well known that the upper surface of blotting paper differs materially from the lower, the one being quite smooth and the other far more open and woolly. Again it should be taken into consideration that the paper, when stacked, presents these surfaces in changing order. Therefore, someone should be charged with arranging the paper with the smooth side up, which is best for printing. It may also be said that upon paper of the kind in question a solid and sharp impression can only be made when a hard, smooth and even cardboard is placed on top of the rubber blanket on the cylinder, and furthermore, the work upon the stone should be etched to a considerable relief. As most of the dust comes from the trimmed edges the paper should be dusted at the edges with a hard brush before being placed on the steam-press for printing: but after all, it will become necessary to scrape the ink rollers several times during the day and to place upon them fresh ink.

Mistakes in Colour Printing.

—It is well known that in all colour work, especially theatrical, show-card, and label work, where but four or five printings are required, the colours are mostly printed in the following order: Yellow, red, black, blue, and if a fifth colour is buff, this comes last. This order is invariably followed, except when it is desired to have in the four or five printings a brilliant green or a good purple. It is impossible to produce a warm brilliant green if yellow is printed before the blue, and it is the same with purple. A blue over red never makes as fine a purple as if the blue were printed first and the red over it. A chrome yellow printed first and a milori blue upon it produces a cold dark green against a warm brilliant green obtained by reversing the order. Milori blue over vermilion gives a dark dirty brown, over crimson it forms a cold, dark-bluish purple ; dark blues, such as Prussian, bronze, and indigo, over vermilion produce an intense black, against which a true black appears decidedly grey. Prussian blue and bronze blue printed over crimson lake appear as a very dark, almost black-bluish purple, while the lake printed over blue gives a true purple. The best and brightest purple obtainable by printing red over blue is secured by cobalt blue and carmine lake. The brightest green is produced by milori blue first and light chrome yellow over it.

Safeguard against Warping of Cuts.

—It is recommended, in printing from original woodcuts, to place a sheet of gutta-percha under the forme. This prevents warping.

How to Improve Bronze-printing.

—Bronze work is very seldom what may be termed as thoroughly satisfactory work. Its failure, as a rule, rests in the inability to fix it firmly on the paper. Of course rolling is the most reliable remedy, but as you may not happen to have a rolling machine, nor the inclination to invest in one, adopt the following method :—Work the forme with gold size and apply the bronze in the usual way ; when the number required is completed, simply take all the rollers off the machine, clean the forme, *but don't disturb it,* and run the sheets through the machine again off the clean forme. The appearance of the work is greatly improved by this process.

Cleansing Formes and Rollers.

—A new invention, calculated to be of considerable service in printing, is the use of oil of camphor for cleaning rollers, type engravings, and lithographic stones. It is preferable to oil of turpentine, being without its unpleasant smell, and containing no greasy substance ; nor does it leave any deposit on the article cleaned. Joined to these advantages, its moderate price will render it a favourite.

Printing Solid Black on Gold.—Care must be taken to obtain a plate of good quality for the gold. One made of metal, and perfectly free from small holes would suffice, but a polished zinc plate, cut to the required size, only costs a small sum, and is the best. Boxwood blocks and other kinds of tint-plates are not so suitable for long numbers, frequent cleaning causing the surfaces to wear rough. The bronzing preparation must be one that dries well when covered with gold, but does not dry so quick as not to allow sufficient time for the bronze to be carefully laid on. A bronzing preparation is recommended to all who have bronze work to do. A large pad of cotton-wool must be used for laying on the bronze, and as the sheets are finished they must be laid perfectly straight until dry. Surplus bronze may then be dusted off with an silk old handkerchief without any fear of scratching the gold surface. Blue-black is the only ink that prints solid black on gold at one impression. Printers will, however, experience some difficulty in getting their work to look sufficiently black and solid without overloading the forme with ink. The whole process is slow and tedious, and one that refuses to be hurried, but the finished effect is so fine that it repays the trouble. The forme, if a fairly solid one, showing large masses of black (similar to that shown in "The British Printer"), should be double-rolled, and the machine run very slowly—so slow indeed as to give the machineman time to peel the sheet from off the forme for every impression. This may be done easily by moving the grippers clear of the sheet and allowing the latter to adhere to the forme ; a quick grip of the sheet with both hands just before the rollers descend will suffice to peel the sheet from off the forme, and most effectually prevent the "lifting" or "tearing-up" of the gold surface, which must inevitably result where grippers are used to pull the sheet away quickly from the forme.

Printing Envelopes.—To prevent the lumpy particles of mucilage on gummed envelopes from "battering" the type, use a heavy piece of blotting paper as a tympan, and when beaten down touch the injured part with a drop of water, which will bring up the impression again.

Set-off Sheets.—A sheet of paper wet with glycerine and used as a tympan-sheet will prevent offsetting. This will be found better than using oiled sheets.

The Use of Tints.—In no department of the typographic art has there been more marked progress during the past five years than in that of colour printing. The origin of letter-press printing in two or more colours seems somewhat obscure. Ancient specimens of press-work are to be found in which ornamented coloured initials, etc., are worked with the text ; but many of these appear to have been the handiwork of the scribe rather than of the printer. Much of the colour printing of the past consists of attempts at showy effects, in bright or positive colours ; but in later years, especially during the past decade, the printer has striven for artistic honours by aiming at harmony of colouring, as well as brilliant effects, so that in this direction great advancement has been made. The principles of artistic colouring are governed by laws, and their observance in the production of artistic effects is necessary on the part of the printer as well as the painter. The latter, however, has the advantage of the printer, because he can apply his colours with the brush in any required degree of intensity, changing and substituting until the eye is satisfied with the effect ; while the printer must, to a great extent, carry his plan in his *mind's eye*, and judge beforehand of the effects of his colours and the harmony of the whole, even before one of them is applied to his work. The necessity for a fair understanding of the laws of colour by the printer will, therefore, be apparent, and he who would be successful in this branch of the art should comprehend at least the simplest principles of harmony and contrast in colour. The employ-ment of tints in the production of fine effects in colour work is becoming a necessary part of the printer's art. By the word tint we mean considerable surfaces of colour, applied, not to the text or lettering, but as an adjunct to or ground-work for the whole or some portion of the lettering. These are employed either to heighten the effects of single lines or groups of lines, or to serve for a background for portions of the work. In either case great care is required to avoid disharmony of colours; the safest tints, on this account, are those in which the primary colours do not appear, such, for instance, as drab, buff, etc. Indeed, it is safe to say that for a ground-work of much surface, a primary colour can very seldom be used with good effect. When the type-work is in several brilliant colours, the tints should always be subdued and " quieting " in their effects. Where a single tint is used as a groundwork for the text, its colour should be made to depend on that of the ink used for the text. If much black, blue, or other dark ink be used, the tint may be " warm," such as buff, orange, pink, or purple. If warm colours are used in the text, a " cool " tint, such as drab, grey, or slate may be employed, always seeking to preserve a balance and harmony in the whole. To be effective, a tint, especially if it be light or pale, should have a well-defined boundary, of a deeper colour than itself. This is usually accomplished by working a tint to fill a prescribed space, defined by rules or border, in which case accuracy of register is an important requisite, as the tint over-running or falling short of the boundary presents an unsightly appearance.

Hint for Centring Lay.—In job work, when an impres-sion is taken on the tympan, and the pressman wishes a sheet to be printed in the centre, he has only to place the right edge of his paper at the right end of the printed line on the tympan, and mark on the sheet at the left end of the same line, and fold the remainder into one half, marking the tympan at the left edge of the sheet to be printed.

Slurring on Platen Presses.—If your forme slurs, try the following remedies, until you strike the right one :—

1.—Take it off from the press and see if it is springy. If it is, and the fault is with the chase, put a strip of cardboard, a pica wide, between top of furniture and chase, all around. This will overcome the tendency to spring.

2.—Examine your tympan ; and here in most cases is the source of the whole trouble. If it is baggy, draw it tight. See if it is not loose at one point and tight at another. If you have several sheets of heavy stock or cardboard under your

F

top sheet, the imprisoned air will cause a slur. This can be overcome by using one thick cardboard in place of the several sheets. If it is warped or uneven, reject it and use a flat sheet. See that the cardboard is not too large. If it is, the tightly drawn tympan will warp it.

3.—Examine the impression studs, and make sure that none of them are loose. It often happens that one stud is drawn back, and is loose, leaving the force of the impression to fall on the three remaining ones. This would very likely cause a spring of the platen, and consequently a slur.

4.—Notice how the grippers strike the platen. They must come down perfectly flat, all points alike, and must not slide upward after they have once touched the platen; neither must they loosen their pressure on the sheet until they leave it entirely. A sliding gripper may move the sheet so as to crowd it against the side gauge, so that the sheet would be baggy when it reached the forme, and a slur would be inevitable.

5.—Try with one gripper alone, then with the other alone.

6.—If the forme will allow it, stretch a rubber band, or two of them (one above and one below the forme), from gripper to gripper. If a two or four page forme, use rubber bands or tape between the pages, making a sort of frisket.

7.—Take a piece of flat manilla paper of good quality, of the right size for a tympan sheet, dampen one side with a sponge quickly, so that the sheet will be dampened through, put it over all the tympan (after first removing sufficient paper to adjust the difference), draw it tight both ways, and let it dry. You will have a tympan as taut as a drumhead. Do not be discouraged if it looks hopelessly wrinkled after first putting it on. It will straighten out when the moisture is gone. If you cannot wait for it to dry, heat a board and lay it on the tympan for a few minutes.

To Overcome Electrical Attraction in Paper.— Take a large type galley and lay it on the delivery table, where the sheets will fall upon it. Run a copper wire from it to the steam pipes, just behind it, and this will carry off all electricity, so that sheets can be straightened easily.

Lubricating Printing Machines.—All machinery should be kept clean, and well oiled every morning ; but newspaper and fast speed machines should be oiled twice a day. It is very important that printers should have an oil for their machinery which is quite free from any tendency to glutinating or clogging qualities. Very often great inconvenience and no slight damage is occasioned by using inferior lubricants.

Save your Make-Ready.—How many pressmen do this ? Indeed, how many pressmen are there who pause to consider how in numerous ways they can save trouble to themselves and time to their employers by a little system and forethought ! The preserving of make-ready sheets of all jobs likely to be done again, whether in type or plate form, is one of these. There are few offices where regrets for neglect of such a precaution have not been expressed. It is a safe rule to keep the make-ready of every type job until the job has been distributed, and even then the pressman should carefully cut out and preserve the make-ready of all cuts or other difficult or tedious work that may be included in such job, before throwing away the rest. As to electro or stereo plate forms, he should invariably keep and file them, for they are useful. Even should the margins be changed, he can easily cut the pages apart and adjust to the new margins by pasting his old make-ready over the pages in their position on his cylinder or platen.

To Work Headings at Press.—Having your paper ruled to the desired pattern, set up your type so that it will register in the compartments of the ruling prepared for it. Then make ready the forme and lift a sheet in, as near register as possible on to the tympan of the press—it is impossible to work headings properly at a cylinder machine—then get some very long darning needles, the longer the better, and stick them firmly into the tympan so that they be flat to the paper. These needles, if stuck as you do pins for laying the sheets to, will guide the register, as they must be so placed as to hide lines of the ruling, both at the off side and bottom side of the tympan.

To Prepare an Overlay.—About the best method of preparing a woodcut overlay is to take three careful proofs on a kindly-surfaced paper, moderately thin ; next cut out slantingly all the light portions of proof No. 1, and set it aside to build upon ; next cut out proof of No. 2 a great deal more, according to the character of light and shade and the judgment of the operator, and paste what remains with as little as will hold exactly in position over the first ; next, cut out such heavy portions of proof No. 3 as may appear judicious, and paste them also in exact position, and the " overlay " is almost completed, but may need a little dressing with a sharp eraser. By this simple process the pressman builds up a finished overlay that so operates on the impression of the engraving as to bring out the shades and manipulate the high-lights of the picture with all the artistic effects originally contemplated by the artist. He is also enabled to " cover " with the least possible amount of ink, which is a leading feature of artistic presswork. But the young pressman must have experience in this particularly nice operation before he can expect to become a proficient. We have described the process ; yet engravings vary so much in their general character that no dogmatic system will apply universally. The student must possess a quick perception of what is demanded by each particular cut, and modify his judgment accordingly. An excellent plan is to keep before him the engraver's proof, and be guided by it in the amount of building up that is expedient.

How to Print with Bronze.—To do good work with bronze powders, it is necessary to select a set of tolerably well seasoned rollers, made of glue and molasses composition ; these should be slightly sponged with water a minute before using. A well-ground size—not too thin nor too stiff—which will work free, and cover with a close and even surface when rolled, should be procured. This should be well distributed before applying it to the forme, and no more carried than is absolutely needed to cover evenly and hold the bronze. When too much of the sizing is used, rough and bad work will result, no matter how thoroughly the bronze powder may be rubbed on. The ink-size and rollers being right, the manner of applying and rubbing in the bronze is next in order. About the very best article to apply the bronze powder to the printed stock is soft and well-picked cotton, as this leaves no scratches across the work. And here comes in the most important and really artistic part of the operation, for of all stages of failure to execute good bronze printing this is the one at which neglect is most apt to show itself. Therefore, take a moderately large piece of the cotton, carefully round off the ragged edges, and dip it in the bronze (slightly shaking back the superfluous powder that may adhere), then apply to the work, gently going over all the printed portions until covered, and the cotton pad slips along more easily ; then increase the pressure and briskness of motion in rubbing over the work for a few seconds, as if polishing, and the desired effect is readily produced. In a word, brilliant and even work is accomplished by the method stated, and the efficacy of a little lively rubbing. Bronze powder is usually put on with too great a pressure ; this causes blue and unsightly marks on the paper or card. Then, again, there are badly compounded bronzes, with apparent good colour, which will produce a like result, no matter how carefully they may be applied. Such a quality should be cast aside, as it will become tarnished in a short time.

The Selection of Colours.—The following table will be found useful in choosing the various tints, inasmuch as by examining them in the order here given, the eye will at once detect the slightest differences of shade. To refresh the eye

Look at Greens	*before choosing*	Reds.
,, Blues	,,	Oranges.
,, Violets	,,	Yellows.
,, Reds	,,	Greens.
,, Oranges	,,	Blues.
,, Yellows	,,	Violets.
,, Tints	,,	Browns.
,, Browns	,,	Tints.

To make Bronze stick.—Those who have trouble in making bronze stick when dry should try a little mucilage or liquid glue in the size. Mix only a little at a time, and wash up frequently or the rollers and disc will become too stiff to work.

Some Hints on Presswork.—Do not try to correct the faults of hurried making-ready by a weak impression, and by carrying an excess of ink to hide the weakness. Excess of ink fouls the rollers, clogs the type, and makes the printed work smear or set off. A good print cannot be had when the impression is so weak that the paper barely touches the ink on the types and is not pressed against the types. There must be force enough to transfer the ink not only on to the paper, but *into* the paper. A firm impression should be had, even if the paper be indented. The amount of impression required will largely depend on the making-ready. With careful making-ready, impression may be light; roughly and hurriedly done, it must be hard. Indentation is evidence of wear of type. The spring and resulting friction of an elastic impression surface is most felt where there is least resistance —at the upper and lower ends of lines of type, where they begin to round off. It follows that the saving of time that may be gained by hurried and rough making-ready must be offset by an increased wear of type. That impression is best for preventing wear of type which is confined to its surface and never laps over its edges. But this perfect surface impression is possible only on a large forme with new type, sound, hard packing, and ample time for making-ready. If types are worn, the indentation of the paper by impression cannot be entirely prevented. Good presswork does not depend entirely upon the press or machine, neither on the workman, nor on the materials. Nor will superiority in any one point compensate for deficiency in another: new type will suffer from a poor roller, and careful making-ready is thrown away if poor ink be used. It is necessary that all the materials shall be good, that they should be adapted to each other and fitly used. A good workman can do much with poor materials, but a neglect to comply with one condition often produces as bad a result as the neglect of all.

Printing Red on Black Paper.—Try printing once with varnish and twice with red if an intense colour is desired. We have not tried it, but the Berlin Typographical Society says it will accomplish its purpose.

Printing in Bronze.—The whole secret of brilliant bronze work is to have the sheets rolled after printing.

Bronze Printing.—In colour printing the gold is worked first, because otherwise the bronze would adhere to the colours and the job would be spoilt—matter in a wrong place you know. If you must work your colours first—and there are cases where it is advisable to do so—the job should be dusted with chalk before the gold is added.

Setting Gauges on a Platen.—Much time is often lost in guesswork. Ascertain the correct margin at the head of job, and mark the correct distance on the platen. Get a straight edge, square it from some line in the forme to the margin mark, and draw a pencil line across platen sheet. Apply your gauges and you cannot help getting them straight first time.

Printing on Glazed Surfaces.—It is well known that printing ink when used on glazed and enamelled paper dries rapidly and pulverizes easily, so that the work is more or less rubbed off. This is due to the fact that the paper absorbs up to a certain point those elements or substances which enter into the composition of the ink and whose function is to bind together the solid elements. In consequence of this absorption the colour or lampblack rests like dust on the enamel and rubs off naturally with great facility. To obviate this inconvenience recourse is had to two different methods: either to modify the paper used or to add some ingredient to the ink which will cause it to adhere better. The latter is the preferable course, for it is the simplest. For printing on glazed or enamelled paper add a varnish rich in resin, such as is used for bronze-work. This causes the colour of the ink to be somewhat deteriorated, but if care is taken there is not much to fear.

Good Ink for Cuts.—It does not pay to try to print cuts or fine letterpress on good paper with a poor quality ink. Use the thick-bodied, short ink for such work, especially on a platen press.

The Printing and Cutting of Labels.—Supposing you wish to trim a circular label quite close to the inclosing rule : Get some steel cutting rule, and with the help of a curving machine first bend a lead so as to fit the rule closely; then bend your cutting-rule to fit around the lead, leaving a very small space (say one-sixteenth of an inch) between the ends; when fixed put your forme on the press ; use a *hard* tympan (Bristol board is good), and after making ready put an overlay of cardboard over your cutting-rule. You will find that the next impression will print your label and at the same time cut it smoothly and neatly from the sheet, except where the space comes between the ends of the rule. This small uncut space serves to pull the sheet from the type, and if the sheets are fed accurately they may be knocked up straight and square and the paper cutter used to separate the narrow uncut space, or a thin chisel may be pushed through the pile. If these directions are followed labels of any shape can be printed with narrow or wide margins, of uniform width all round, and without waste of time. On odd shapes, where the cutting-rule has to be fitted in sections, it is best to use the soldering iron to keep them in place. The same device may be used to cut cards to odd shapes by fitting the rule to the shape desired, and running cards through the press without rollers.

Cleaning Machinery.—Take half an ounce of camphor, dissolve in one pound of melted lard; take off the scum and mix in as much fine blacklead as will give it an iron colour. Clean the machinery and smear with this mixture. After twenty-four hours rub clean with a soft linen cloth. It will keep clean for months under ordinary circumstances.

Barking the Knuckles.—In washing small job formes with benzine after taking proofs, printers find it provokingly easy to scrape the skin off the knuckles while driving the small brush generally used across the forme. One who has tried says that the cause of knuckle barking is the smallness of the brush, and that after taking into use an ordinary boot-blacking brush he saved his skin.

Blending of Colours.—The laws of harmonious colouring are a necessary part of the knowledge of printers in colours. A few contrasts are :—

Black and warm brown.	Violet and pale green.
Violet and light rose colour.	Deep blue and golden brown.
Chocolate and bright blue.	Deep red and grey.
Maroon and warm green.	Deep blue and pink.
Chocolate and pea green.	Maroon and deep blue.
Claret and buff.	Black and warm green.

Method of Printing in Colours.—A new method of colour-printing consists in taking any desired suitable pigments, preferably aniline, and dissolving them in any suitable solvent—such as water, alcohol, glycerine, oil, or any desirable mixture of either—and so prepared that the colours will not too readily set. Sheets or rolls of tissue-paper, silk, or other suitable fabric are then taken and saturated with the prepared pigment, different parts of any given sheet being saturated with different colours, and the pigment being applied in strips, spots, or according to any desired design. The paper will then usually need to be dried, or partially dried, to enable it to be handled and prevent smearing, and insure a clean, sharp impression. The pigment may be applied by means of ordinary printing rollers, such as are used in calico-printing ; or the paper may be impregnated by means of pens, pencils, or brushes, such as are used in preparing ruled paper ; or, instead of a single sheet saturated with more than one colour, tissue-paper or fabric may be first impregnated with one colour, using it as a base, and then strips or designs of other paper or silk, each saturated with a different coloured pigment, may be laid over or secured to it, the whole being then dried as far as necessary, or the many-coloured sheets may be made of pieces connected at their edges, either before or after saturation with pigment, and then dried. One of the coloured sheets, prepared in either of the ways indicated and with any desired design, is placed in contact with the paper to be printed and between the print-paper and the type, electrotype or woodcut, and the whole is then passed under the cylinder or platen of the press, as in ordinary printing, and the impression in two or more colours in any desired design is thus made without the use of the ordinary inking process; or the face of the type is first rolled with ink, so as to make a base colour, and then there is laid over it, on the face of the type, coloured strips, and thus any desired number of colours may be printed at one impression.

About Inks.—The management of inks seems little under-stood by many printers. Printing ink is substantially a paint triturated to extreme fineness. There are occasions, of course, when the least amount of colour that can thus be put on is sufficient, but it generally needs more. Especially in handbills and posters more is required. The first requisite in this case is that they shall catch the eye quickly, which cannot be done by hair-line faces or small quantities of ink. They should be charged with colour. Principal lines should have more impression than weaker ones, and this is generally better accomplished by underlays than overlays, for not only is the impression stronger, but the line will thus take more ink. The more slowly the impression is made the blacker the line will appear, as the ink has then time to pene-trate. It is well sometimes, when extra solidity of colour is required, to run a good piece of work through a second time. House painters do not finish a house at once, but lay on one coat after another until the requisite intensity of colour is obtained. Especially should this precaution be followed in pale or weak colours, such as the various yellows. One great reason why this hue is hardly ever used by printers, except through bronzing, is that it always looks pale and ineffective on paper, and is lost in artificial light. The colour, in its various modifications with red and black, is very effective, as can be seen by looking at the leaves of trees in autumn, which are compounds of green, brown, red and yellow, the first soon disappearing, and brown being the last.

Printing of Coloured Inks.—The freshness is taken out of bright coloured inks by their being distributed on the iron surface. Slate or marble is preferable.

Hints on Using Inks.—All ink, when opened, should be lifted—especially in thick and dryer inks—from the top, keeping the surface as smooth as possible, otherwise the side portions will " skin," and be apt to mix with the rest of the ink, spoiling the whole. All cans should be kept carefully covered to prevent " skinning," and all scrapings from table and boxes be kept separate for poster or coarse work. If water gets mixed with ink, it will make it roll out badly and work specky. Any skin formed should always be carefully removed, to prevent its mixing with the undried ink.

Hard Packing and Fine Printing.—Book-printers gave up damp paper reluctantly. For the new method of printing dry compelled them to give up the woollen blanket which had been used between the paper and the pressing surface as the equalizer of impression ever since the invention of printing. That such an elastic medium was needed when types where old or of unequal height, or when the pressed and pressing surface of the press could not be kept in true parallel, needs no explanation ; but the use of an elastic printing-surface was continued long after these faults had been corrected. The soft blanket, or the India-rubber cloth, often used in place of it, made an uncertain impression, which either thickened the fine lines of a cut, or made them ragged and spotty. It would have been useless to get smooth paper if the pressing-surface behind the paper could be made uneven. To get a pure impression it was necessary to resort not only to the engravers' method of proving on dry paper, but to his method of proving with a hard, inelastic pressing-surface. A substance was needed which could be pressed with great force, without making indentation, on the surface of the cut, and on the surface only. This substance was found in mill-glazed " press-board," a thin, tough card, harder than wood, and smooth as glass, which enabled the pressman to produce prints with the pure, clean lines of the engraver's proof. Old-fashioned pressmen prophesied that the hard printing-surface would soon crush type and cuts ; but experience has proved that, when skilfully done, this hard impression wears types and cuts less than the elastic blanket.

Hints for Selecting Paper to Suit Ink.—Printing ink appears, when on white paper, blacker and colder than on tinted paper; while on yellow or tinted paper it appears pale and without density. For taking printing ink most perfectly, a paper should be chosen that is free from wood in its composition, and, at the same time, one that is not too strongly glazed. Wood paper is said to injure the ink through the nature of its composition. Its materials are very absorbent of light and air, and its ingredients go badly with colour. Pale glazed or enamelled paper, on the other hand, brings out colour brilliantly.

Hints for Quickly Drying Printed Work.—It may not be generally known that ink will dry very quickly on paper damped with glycerine water. Posters with large and full-faced types will dry in a quarter of an hour, whilst the drying process, when the printing has been done on paper simply wetted in the ordinary way, will require hours.

Printing Red Ink and Electrotypes.—The following plan answers well for temporary purposes:—Take one ounce of prepared gold size, and a quarter of an ounce of the "lake-brilliant" of Cornelissen, and grind well together with a muller; roll the electro with this preparation and let it stand for twelve hours, when it will be found as hard as stone, and the vermilion may be printed from the plate without the least injury to its brilliancy.

New Colour Printing Machine.—A German firm has patented an improvement in platen machines with oscillating table, by which it is possible to print in two colours without changing the inking apparatus. The present platen or the forme receives a second platen or a frame which may be revolved or displaced in the centre in an axis of the table. A suitable arrangement turns this second plate before closing the platen, or before it is pressed against the forme, by 90 degrees, and the same happens on the retirement of the platen. The inking apparatus is divided into two parts. If only one colour is to be printed, the two parts are pushed together, and thus form a single inking apparatus. The same is the case with the inking rollers.

A Set-off Paper.—Pressmen should remember that paper saturated with benzine is as good, or better, and much cleaner than oiled paper, to avoid a " set off," when work has to be printed on both sides.

To Print Process Blocks.—Many printers object to use process blocks, as they become clogged with ink and require frequently to be cleaned. This is one of their defects, and must be submitted to whenever there is anything more than an outline drawing. Much of the difficulty in printing can be obviated by using nothing but hard paper in the over-lays, care in rolling, and a firm paper to be printed on. The fibres will not then sink into the plate and the ink spread over the portion in relief. Extra care will obviate nearly all trouble from this source.

The Creasing of Sheets on Cylinder Machines.—A correspondent wrote on this subject to one of the London trade journals and was answered thus :—

This query has been dealt with in these columns before. Precisely the same troubles visit printers, no matter where they may be, and if our correspondent had referred back, he might have discovered the information he now seeks given to another printer whose troubles in the same direction were prior to his own. The paper creases. Of course it does. What sort of paper is it ? Is it worked dry or wet ? Was it too wet ? Is it worked with a blanket on the cylinder, or paper, or that oiled canvas, table-cover sort of material ? If we had to work a similar sheet we should get the machine quite clean and sufficiently well down on its bearings before beginning to make ready. Then we should put some sheets of paper on the cylinder, just enough to bring it at its printing surface to a level with the extra thickness of iron which is on the ends of the cylinders. These sheets we should paste at both edges, and put them in their place damp. When they had dried they should be as smooth as glass and as tight as a drum. The forme should be laid on and put, if possible, into its place at once, for if there are several impressions of it the sheets get indented and spoiled. The impression should be very light to start with, and the job should be worked as light as possible. Very little patching should be done, and a sheet might be placed over all before starting, if by so doing the pressure will not become too great. A glazed board from the warehouse might be put on the cylinder if necessary, so as to prevent the sheet puffing, or the canvas we have referred to might be used instead of a blanket. If none of these things produce the sought-for result, a " bolster bearer "—that is, a rolled-up piece of paper—may be stuck on the cylinder below the margin of the page. If this fails no amount of description will help our correspondent. Nothing short of positive assistance personally and practically rendered can be of the slightest use. We have endeavoured to show how not to get this crease, but no fixed rule exists, and such matters have to be met with care, and conquered with dodges, wrinkles, and the like, which can hardly be conveyed at the right moment by means of a printed discourse. As a proof that pressure is the principal cause of creasing, let our correspondent try, when he has worked off the next number of his journal, whether he can make a crease by adding more pressure, or by loosening the back edges of the cylinder sheets. He will get his crease immediately by so doing, and the inference is that by removing the cause the effect should also cease.

Hints on Colour.—The following hints to letterpress colour printers will be found invaluable:—Yellow and carmine or deep red produces scarlet or vermilion ; carmine and blue produces deep lilac, violet, and purple ; carmine, yellow, and black produces a rich brown ; yellow and black, a bronze green ; yellow, blue, and black, deep green ; carmine and white, pink of any shade ; ultramarine, white, and carmine, deep tones of lilac ; violet and white, pale lilac or lavender ; cobalt and white, lively pale blue ; and Chinese blue, deep bronze blue, chrome, pale lemon, any tone of emerald green. Amber is made from pale yellow, chrome, and carmine. Red brown is made from burnt umber and scarlet lake. Light brown is made from burnt sienna shading with lake. Blue and black are made from deep blue or deep black.

Salmon is made from burnt sienna and orange, shading with white.

TWO COLOURS WHICH HARMONIZE WELL.

Scarlet red and deep green.   Light blue and deep red. Orange and violet.   Yellow and blue.   Black and light green. Dark and light blue.   Carmine and emerald.

THREE COLOURS.

Red, yellow, and blue.   Orange, black, and light blue. Light salmon, dark green, and scarlet.   Brown, light orange, and purple.   Dark brown, orange yellow, and blue.   Crimson lake, greenish yellow, and black.

FOUR COLOURS.

Black, green, dark red, and sienna.   Scarlet, dark green, lavender, and black.   Ultramarine or cobalt blue, vermilion, bronze green, and lilac.   Sienna, blue, red, and black.

Printing on Tin.—The impression is first made from the stone on to a rubber roller, and from this roller is transferred off again upon the tinplate.   There has lately been invented a steam press working on this system.   The rubber has the quality of taking a much sharper impression than any kind of paper, and by its elasticity prints smooth and solid on a hard surface even if the latter is uneven.

New Ticket Printing Machine.—A New York printer has invented a very ingenious combination of printing-press and several other similar devices, whereby he is enabled to print ballots and bunch them at the rate of three millions an hour.   The paper is run in full sheets, but it is cut into strips immediately after leaving the cylinder, whence it goes through the folder in as many slips as the full tickets require.   Then the strips are cut into pieces the size of a ballot, and finally encased in a pasted envelope, in form like a flat cigarette. The machine is comparatively simple in construction, and is designed also to print telegraph blanks and do them up in pads at the rate of 10,000 to 12,000 per hour.

A new Perfecting Machine has been patented in Germany. In this machine two simple rotary motions have been combined, so that only one motor moves the two formes and the two printing cylinders. The sheet, after being printed on one side, is turned by a turning drum during the return of the carriage, and then printed on the other side during its next forward movement. Both the prime impression and the reprint are consequently effected during one passage of the carriage on two sheets. It follows from this arrangement that the two printing cylinders are not continuously revolving, but are arrested during the return of the carriage, as in ordinary machines, by an intercepting fork. The sheets may consequently, during this pause, be pressed against points. The machine delivers for each turn a perfected sheet, with the exception of the first and the last turns. In the former the reprint cylinder revolves without printing ; in the latter the prime cylinder turns without doing work.

To Test Leather Belting.—A simple method of determining the value of leather belting consists in placing a cutting of the material about 0.3 of an inch in thickness in strong vinegar. If the leather has been thoroughly acted upon by the tannin, and is hence of good quality, it will remain, even for months, immersed without alteration, simply becoming a little darker in colour. But, on the contrary, if not well impregnated by the tannin, the fibres will quickly swell, and, after a short period, the leather becomes transformed into a gelatinous mass.

Cleanliness in the Press-Room.—A correspondent writing to an American journal on the subject of cleanliness, says, " Thinking the matter over I was planning how the press-room could be kept clean, and the thought came, how often when a person happens into a press-room he finds it dirty, everything every way, and, worst of all, the presses covered with oil, and under each a veritable oil well. Then the pressman tells you that they have been very busy lately and have not had a minute to clean up. I venture to state that this.

pressman (and there are a good many such as he) will give a similar excuse if you drop in just after they have had a short run or dull time.   Now, a plan I suggest is, each time you go to oil up a press, take a piece of waste or woollen rag, and when the oil-hole is properly filled (bear in mind the hole was not made to hold a pint), wipe off what little runs on the outside, at the same time give a quick rub to parts close by, then whatever has worked out during the previous run will be cleaned up and you will find very little has gone on to the floor, or, better still, if you can have it, zinc under each press. If this is done by the pressman or feeder, every time you oil up, much of the dirt, saying nothing of the soiled jobs, which you see in many press-rooms, will be cleaned up and it can be done with hardly any loss of time, what some call it, but I think done with a saving of time."

Cleaning Rubber Blankets.—We learn that the use

of turpentine in removing grease and colour from rubber blankets, is increasing to such an extent, that we desire to make a few suggestions as to its use and effect.   The quantity used should be as small as possible, and great care taken that it is thoroughly dried out before the blanket is used in printing.   Otherwise, as turpentine softens the rubber face, the blanket will be injured by the pressure of the cylinder causing wrinkles to appear upon the face.   It is preferable to clean the blanket after work at night, thereby giving ample time for the turpentine to dry out, rather than in the day-time when the press is in use.   The use of ammonia, as a substitute for turpentine, is strongly recommended, with less chance of damage to the blanket.

Dirty Tapes on Machine.—A great drawback in rotary

machines is the liability of the tapes to get dirty.   In the first place, the use of ink containing a sufficient proportion of varnish is recommended.   The latter may be dearer, but it is more economical.   In the next place, thin tapes are suggested ; if they are not drawn too tightly they last quite as long as thick ones.

G

Preservative for Belting.—New belts should have enough dressing in them to last several months, unless they are getting very hard treatment. Castor oil is a good preservative dressing.

Cotton Belting.—When first put on they require "taking up" once or twice more than leather. The stretch is approximately six per cent. as against four per cent. in leather, but once well at work they give less trouble, as there is but one joint to look after. If put on really tight enough, this stretching is diminished to a great extent. Users are afraid to overstrain the belt, but experiments made as to tensile strain show the impossibility of breaking a belt under fair conditions.

Erecting the Columbian Press.—Place the feet on the staple and raise it upon them; then place the bar-handle in, with the bolt belonging to it; put the principal lever into its place, and then the bolt which connects it to the staple; then the angular or crooked part, which has one square and three round holes, through it, into the mortice, which is in the projecting part of the long side of the staple, and place in the bolt that attaches it to the staple. In the extreme edges of the heads of the two before-mentioned bolts you will observe marks, and corresponding marks over the holes through which they pass; put the bolts in so that these marks meet together and correspond, and so on, until you have all the remaining parts in their respective places. The four screws for the platen, which have heads on one side, are intended to attach the platen to the piston, which being placed in their proper places, are secured by the four small blocks of iron which accompany them. To increase the power, turn the nut in the rod so as to shorten it, and to decrease it, turn it the reverse way. By the nut on the iron screw, which connects the main and top counterpoise levers, you may regulate the rise and fall of the platen, so as to clear the head-bands of the tympans, which is done by screwing the iron nut up as far as is necessary. In adjusting the platen so as to approach the forme exactly parallel, you must, after hanging on the

platen and having a forme on the table, square it to your
tympans, then make a pull, and hold the bar-handle home
until someone else screws the four platen bolts to an equal
tightness. The small holes which communicate with the
different bolts require a small quantity of machine oil occasionally. The impression may be increased by putting thin
pieces of tin or sheet iron, cut to the size of the plate of iron
which lies between the platen and the piston, secured by the
four screws on the top of the platen, and placing it under
the piston; you can then readily tell whether everything is in
its proper place, by the perfect ease with which the bar-
handle acts. As will be gathered, the impression is obtained
by means of levers, somewhat on the plan of those used by
Earl Stanhope in his press. The power is given by the heavy
cross-beam at the head of the press, set in motion by pulling
the bar-handle across, which acts on the horizontal rod
attached, and also brings the elbow into play, great power
being thus obtained. The top cross-rod, on which the eagle
is placed, is the counter-weight, which falls back into its
original position—having been raised in the act of impression—
when the recoil of the bar-handle has taken place. These
presses are made as large as double royal.

Erecting the Albion Press.—Place the feet on the
staple and raise it on them; then place the spring and box
on the top of the staple, dropping in the long loop bolt, which
is connected with it, into the long hole in the staple; then
connect the piston by passing the round bolt through the
hole in the staple, and fasten with pin and washer; put the
bar-handle in its place with bolt, tightening it so as to allow
the bar-handle to be free; then attach on, with the four
screws, the slides or guide-pieces to piston; then place the
chill into the piston, also the tumbler or wedge-shaped piece,
taking care that the bright or numbered side is towards the
bar-handle; then connect the chill with the bolt in the
handle, screw up the nut or top of the spring-box sufficiently to draw back the bar, so as to keep all parts in their
places. The wedge and brass guard in front of piston are
intended to regulate the pressure. The other parts of this

press may be fixed in the same manner as the Columbian. These presses are also made in very large sizes. The power is obtained by means of levers, which act on an inclined piece of steel called a chill; by pulling the bar-handle across, this chill is brought from the sloping into a vertical position at the precise moment of impression. On the bar-handle being allowed to go back to its original position, the chill resumes its former inclination, and the platen is raised from the surface of the type by the recoil of a spring contained in the box at the head of the press; this then allows the forme to be run out, rolled, and run in again for successive impressions, the sheet as printed being first removed, and another laid on its place.

Packing Rollers on a Machine.—This is sometimes necessary to avoid the " wiping " of the roller on the extreme edges of the type, which causes an excess of ink on the part where the rollers pass over any opening between the pages of the forme. In order to get over this difficulty, thick card, or even pieces of leather, may be used as packing; the length should be a little more than the opening to which they are placed opposite. Let the extreme edges be bevelled off so that the rollers will run over without jumping. The exact height will be determined by experience, but generally a sixth or an eighth of an inch in thickness is sufficient.

Finding the Pitch on a Cylinder Machine.—The best plan is to daub a little ink on one end of the cylinder opposite the grippers, and turn the coffin under till this mark is transferred to the impression bearer. A gauge should then be cut for future use, and the formes of works of varying margins may be adjusted to the exact position at once without fear of accident.

On the Mixture of Colours.—It is a common definition of *white*, that it is the presence of all the colours, and of *black* that it is the absence of all colour. It is true, theoretically, that the mixture or combination of the colours of the prismatic spectrum, by means of a lens or concave mirror, produces a ray of white light; but when we mix pigments or inks, representing those colours, taken as pure as we can possibly obtain them, the mixture is not white, but grey or black, according to their intensity, etc. :

For every blue ink contains also either red or yellow;

Every red ink contains also either blue or yellow;

Every yellow ink contains also either blue or red;

And although, as we have said, the union of the blue, red, and yellow of the spectrum produces *white*, the union of blue, red, and yellow inks produces grey or *black* according to proportion.

If we had inks that were in colour as pure as those of the spectrum, their mixture would also yield pure colours.

Ultramarine is the only ink that approaches a prismatic colour in its purity, but even that has a slight tinge of red in its composition, causing it to appear violet.

We can take gamboge as the representative of pure yellow, carmine as that of red, and Prussian blue as that of blue.

In mixing inks to obtain pure secondary colours, we shall obtain a better result if we select such as are free from the colour not essential to the compound. Thus, to obtain a pure green, which consists of blue and yellow only, we must take a blue tinged with yellow rather than with red, and a yellow tinged with blue rather than with red; if we took either of those inks tinged with red, a quantity of black would be formed by its mixture with the two other primaries, and the green would be tarnished or broken. So long as pure blue and yellow are mixed together, in varying proportions, but without the addition of the other primary colour (red), the resulting compound colour, green, remains a pure colour. Such is the theory, and the practical result is the same if the inks we select to form the mixture are both free from the third primary.

When the three primary inks are mixed together in equal strength and proportions, the resulting compound is black. But if they are mixed in unequal strength and proportions, the mixture is grey, coloured by the primary or the secondary in excess in the compound.

Normal grey is formed by mixing a black with a white ink in varying proportions, producing various tones of grey.

By adding a primary or a secondary to normal grey, we produce a coloured grey.

There are as many classes of grey as there are primary and secondary colours, and as many hues of grey as there are hues of these pure colours.   What are commonly called Tertiaries, are, in fact, coloured greys : thus, russet, is red-grey, citrine is yellow-grey, olive is blue-grey.

If the primaries are mixed in unequal proportions, or are of different intensities, the mixture is a grey.

If the blue is in excess, the mixture is a blue-grey.

If the red is in excess, the mixture is a yellow-grey.

If the blue and the red are in excess, the mixture is a violet-grey.

If the blue and the yellow are in excess, the mixture is a green-grey.

If the yellow and the red are in excess, the mixture is an orange-grey.

When two secondaries are mixed together, the grey that results is coloured by the primary which enters into the composition of both secondaries, thus :

In mixing green with violet, the grey is coloured by blue, that being the primary in excess.

Green consists of blue and yellow.
Violet consists of red and blue.
The compound contains as much blue as red or yellow.

In mixing green with orange, the grey is coloured by yellow, that being the primary in excess.

Orange consists of red and yellow.
Violet consists of red and blue.
The compound contains twice as much red as blue or yellow.

Printing on Xylonite.—W. L. Brown & Co., New York, announce that they are now prepared to furnish the trade with indelible inks for printing on Xylonite.   These inks are said to be so prepared that they can be successfully used by any printer on an ordinary printing-press.   There are twelve distinct colours, which are put up in one-pound cans.   Xylonite ivory being a very attractive and durable material for advertising and display cards, it is believed that these inks will prove of great advantage to the trade. Heretofore Xylonite has been printed upon by an expensive process, heat and pressure being necessary in connection with steel plates; but by the use of the new indelible ink this is overcome, and the Xylonite card is placed within the reach of printers having only an ordinary outfit.

A Hint for Printing on Xylonite Cards.—In printing on these plates, the sheets are first slightly damped with alcohol, which dissolves or softens the surface and allows the ink to soak into the celluloid.

Sizes of Platens of Hand-Presses.—Presses are made in various sizes, and classified, commencing with the smaller, as card, quarto, folio, and broadside, corresponding with the different sizes of paper used in printing. The size of platen determines the classification, and the same conditions apply to both kinds of presses' mentioned. Card or quarto presses measure something less than the smallest dimensions now given.

Name of Press.	*Size of Platen.*
Foolscap folio	$15 \times 9\frac{1}{4}$ inches.
Post ,, ...	16×11 ,,
Demy ,, ...	18×12 ,,
Foolscap broadside	$19 \times 14\frac{1}{4}$,,
Crown ,, ...	21×16 ,,
Demy ,, ...	24×18 ,,
Royal ,, ...	$26 \times 20\frac{1}{2}$,,
Super-royal ,, ...	29×21 ,,
Double crown,, ...	$34 \times 22\frac{1}{2}$,,
Double demy ,, ...	36×23 ,,
Double royal ,, ...	40×25 ,,

Cylinder and Platen Machines.—The difficulty of keeping good register on the perfecting machines is doubtless caused by the platen machines being driven from the same shafting. The platens should always be worked separately.

ROLLERS.

ROLLER Composition.—Owing to the introduction in recent years of many patent and ready-mixed compositions, very little is known in most printing offices as to the ingredients of composition rollers. One of the principal desiderata in a composition is its non-liability to be affected by change of temperature ; for this purpose the use of glycerine is very valuable, as it is little affected by heat, cold, or frost (it never freezes), while retaining moisture much better than the old treacle and glue compositions. The addition of glycerine, however, necessitates different treatment in regard to cleansing. Water should not be used on a roller containing glycerine or gelatine. The tendency of heat on rollers is to soften them, and cold to harden them; therefore for cold weather the ingredient which gives softness to rollers should be in larger proportion than in hot weather. If the ordinary recipe were treacle eight parts and glue four parts, for cold weather it would be best to give three parts of glue and nine parts of treacle. These proportions would depend very largely, however, on the quality of the glue, experience teaching that this varies in a large degree. Another good recipe for cold weather would be—glue ten parts, sugar ten parts, and glycerine twelve parts. The glycerine will offer strong resistance to frost and cold.

Best Period for Making Rollers.—The months of April and May are the best to get rollers for summer use. Don't wait until hot weather is on you. It cannot be expected that winter-made rollers will work well in hot or muggy weather.

Damp or Greasy Rollers.—These may be known by their printing a dull, dirty grey instead of a proper black. If new, wash them in turpentine; if old, in lye. A far better plan is to smother them in common ink-scrape, and sheet them ; this is always effectual with greasy rollers. If a damp roller does not recover with this treatment, you should hang it up in a warm room until it does.

Remedies for Damp and Dry Rollers.—Sometimes pressmen find fault with their rollers, and there may be reason for grumbling. If a remedy is wanted on the spot, one or two suggestions noted lately may prove useful. If a roller is affected by moisture in the atmosphere, wash it with common alcohol, which evaporates the moisture very quickly. If the roller is too dry, take 100 parts of glycerine, 10 parts of ammonia liquor, 40 parts of old beer, which has turned sour; mix it well together, and wash the rollers for about five to ten minutes, and even longer. The ammonia re-opens the pores of the surface for the glycerine ; the beer, by drying up, creates a compound which adheres strongly to the surface of the roller.

Cleansing of Rollers, etc.—For some time past Germans have been making use of oil of camphor for cleaning rollers, type, woodcuts, machines, etc., in preference to turpentine, petroleum, and benzine. The qualities which recommend it as a cleansing agent are not to be overlooked. Firstly, it is cheaper than the other liquids ; secondly, it is healthy, and acts as a disinfectant in the printing office ; thirdly, it is as efficacious and as rapid in its action as any of the products employed up to the present ; and fourthly, it contains no fatty matter, and leaves no deposit.

A Method of Renewing Printers' Rollers.—A Berlinese has invented a cheap and simple way of renewing old, worn-out rollers. The rollers are first cleaned with acetic acid, then revolved in a solution of gelatine and glycerine, which fills up the holes and forms a thin coat on the top of the rollers. This dries while they are revolving, and makes the rollers equal to new.

Cracked or Cut Rollers.—These are almost incapable of being mended. If they are cut by working rule formes they may be carefully " flared " with a piece of lighted paper, and hung for a day or two to recover themselves. If split or cracked by accident, let the flame of a lighted candle into the fissure and close carefully with the fingers ; then let the roller be hung up to recover itself. The fingers must be wet, to prevent the composition sticking to them. Caution is necessary not to use a cut roller for colour-work, as the old black ink will be sure to ooze out of some of the apertures. When working at press always cut round those parts of the roller-handles which rest upon the ink-table before commencing colour-work, or the black ink which adheres to them will deteriorate the colour.

Recipe for Softening Rollers, etc.—The following formula will be found invaluable in the press room :—

Boiled linseed oil	$2\frac{1}{2}$ oz.
Balsam fir	$1\frac{1}{2}$ oz.
Demar varnish	2 oz.
Balsam copaiba	3 oz.
Simple syrup	$2\frac{1}{2}$ drams.
Mucilage (gum arabic)	2 drams.
Pulverized alum	$\frac{1}{2}$ dram.

Shake well and let stand for several hours before using. A little of it softens up sticky rollers or ink, reduces ink without affecting its colour, makes cold-weather printing possible, is just the thing to mix various colours with in making tints, and is in every respect more useful than the ordinary printers' varnish. Try it and see if it is not so. The formula will make about a quart.

Roller Fretting.—Put a roller on the slab in the proper place in the carriage, see what you can get between the end of the stock and the roller carriage (try a nonpareil first, or a pica), and what you can get in will be the proper amount to put in under the carriage ; it will then lift the roller parallel to the slab. The real reason of "fretting" is because the carriage lets the roller drop *too* low.

Good *v*. Bad Rollers.—On the pressing necessity of good rollers for good printing a book might be written. Be the type ever so new and bright, the machinery of the best and most approved patents, and the ink up to the mark, without a good roller all the labour and industry will be greatly marred. The roller, black as it is, dull as it is, unpleasant to handle as it may be, is the secret that, once learned, makes it an easy task for a printer—with proper care in other directions, of course—to print well. On the maker depends the quality of the roller, and on the pressman depends the duty of keeping it in order.

Warming Rollers.—If rollers have become too cold, place them in warm rooms, not near the fire, until they recover themselves. When they begin to work, place a candle or gas jet under the ink-table—if an iron one, the flame about a foot below the iron—and occasionally vary the position of the light. The table must not be made hot, and the heat applied must be imperceptibly small, as the object is only to remove dampness and the rigidity of the cold.

Rollers out of Use.—Rollers put away in an upright position, and allowed to remain idle for a week or longer, are observed to have a smaller circumference near the ends than at the middle. To avoid this it is suggested that the rollers should be placed in boxes lying lengthwise, with a bearing at both ends in the sockets provided for the purpose.

The Care of Rollers.—Rollers may be kept with a fair working surface if put away with very oily ink on them. If you put them away with ink that has much varnish or dryers in it, you make short work of your rollers; for they then harden and crack along the face, and the ink cakes upon them. The "hardening effect of the atmosphere" is as nothing to the contracting and splitting effect of dryers that are allowed to cake on the roller. Even the too frequent use of benzine, strange as it may seem to many, has the effect of gradually shrinking a roller. Lye is destructive to the face of the rollers when used by inexperienced or careless persons. It is injurious in the best of hands when frequently used.

Put your rollers away well coated with thick oil that is free from acid or alkali, oiling the ends even more carefully than the face, and you will have no trouble. The whole theory of roller treatment, however, is the prevention of the escape of the moisture necessary to retain elasticity, tackiness, and the other qualities that go to make a perfect inking implement. A sponge-pan in the roller-box is the most valuable aid to the longevity and working power of a roller. As to putting away inky rollers, be wary of so doing, unless you are familiar with the quality of the ink you are using. If this is of the quick-drying job variety, we strongly advise washing off after using; and see that the rollers take their drink through their natural organ, the pores, from the humid atmosphere of the roller-box or closet, before they are used again.

Arched Rollers.—The arching of rollers is generally caused by the roller becoming dried or hard. When it takes place, and the outside edges bear off the centre of the roller, cut a strip of the composition off each end with a sharp knife.

Flaring a Roller.—This operation is both dangerous to the roller and unsatisfactory in its results, but nevertheless we give particulars how to proceed with it, as it is a very favourite device with London pressmen, especially in small offices. It is as follows:—Get a sheet or two of waste printing paper, make it up into a loose torch, and when lighted flare the roller all over, just sufficient to add a new face to the roller without melting the composition. This plan will, if successfully executed, close up the fissures caused by the cutting of brass rule.

Washing Rollers.—Rollers must not be washed with lye until they are beginning to wear somewhat. Smothering in common ink and scraping are recommended in preference. Old enamelled-faced rollers may be washed with turpentine with great advantage, as it makes them sticky, which pro- duces a slight temporary suction in the roller. Washing with turpentine will soon spoil a good roller, the stickiness produced becoming a dryer or coating on the face of the roller.

Special Lye for Washing Rollers.—The following recipe has been found successful:—2 lbs. washing soda (bruised), 2 lbs. brown unslacked lime, and 2 ozs. common table salt mixed in three gallons of soft water, stirred. When settled, pour off the liquor, and throw the sediments away. When washing the rollers, sponge dry, with this lye only. It is ready for use in an hour.

Cooling Rollers.—As ice-chests are now very common, it may be well to warn the printer on no account to put a roller in such a receptacle, or it will soon become frostbitten and utterly useless. It might be thought that when the roller is almost separating from the stock in hot weather almost any mode of cooling it would be desirable. The sudden chill to the surface is, however, quite enough to spoil the roller before the cold has time to penetrate sufficiently deep to harden the roller. If unworkable, owing to the excessive heat of the atmosphere, hang the roller all night in a cool, dry cellar, or in any cool place where a good current of air can get to it. If it is a small roller, swing it well to and fro for a quarter of an hour before using it.

A Patent Roller Composition.—A special treatment for composition rollers, having the advantage, it is claimed, of keeping the ink clean and free from dross, has been patented. Before the composition (which is of the same kind as that commonly used) is applied to the stock, certain proportions of either one or more of the following chemicals in solution are mixed with it:—Bichromate of potash, bichromate of ammonia, chrome alum, or tannin. The stock is coated, and then exposed to daylight, or strong artificial light, which exposure to the light renders the composition insoluble. The composition is then ready to receive an oil varnish. It is further claimed that the rollers are rendered very smooth, and will ink much quicker than ordinary rollers, and that machines can be run at greater speeds.

Rollers and Blue Ink.—To avoid the streaky or small-poxy appearance so often seen in solid blue surfaces, more especially when ultramarines are used, see that the rollers are soft, and do not carry too much colour.

Difficult Rollers.—If printers' thin varnish be added to the ink in small quantities it very frequently allows the roller to work, although with a pale, greyish appearance. The ink we now suppose is good, and the roller unruly. This expedient should only be resorted to on emergencies, although it is a favourite wrinkle with some pressmen. If you have a very hard roller and wish to work a light job, such as a circular, spit four or five times on the ink-table and distribute with a small job roller, holding it as you would a mallet, and well hammer the table with it, distributing it at the same time; that is, each time the roller strikes the table draw it along the table to the edge. Five or ten minutes bestowed upon the roller in this manner will cause it to work off a short numbered job quite satisfactorily.

Patent Felt Rollers.—This invention consists of a roller or cylinder of wood or metal, the same being covered with flannel, and over this prepared felt. The backing, where such is used, may be flannel or cloth, wound about the roller, the protecting layer being of oil-silk, oil-cloth, or water-proofed fabric. The felt layer is saturated with a mixture composed of substantially equal parts of refined beef tallow, copal-lac and varnish, the said substances having been well mixed and stirred in an equal quantity of turpentine oil. The roller when dry is ground and smoothed. The roller is claimed to possess several advantages in transferring ink or colour to designs, type, etc., and in lithographic and zinco-graphic printing. It is stated that the felt cylinder does not expand like leather, and is not altered by impregnation with colour.

To Dry or Warm a Roller at a Short Notice.—Hold it a few feet from a moderate fire and keep turning the roller on its axle for five minutes. Or, take some sheets of waste paper and make them as hot as you can, wrapping the roller in them, one after the other. If a new roller be required from a roller-maker, always send a blanket with the boy, to wrap it in. If to be sent by train, rollers should be suspended in a box.

Too New Rollers.—Coat the roller by distribution with balsam copaibæ and let it hang for two hours ; after which, scrape it. This evil-smelling drug is also very useful if mixed with black or coloured inks when they do not work satisfactorily.

Facts about Rollers.—The setting of a roller, especially on a cylinder press, requires care and judgment. Rollers cast from re-casting composition never shrink. Roller trucks should be one-sixteenth of an inch less in diameter than the roller. Glue and molasses rollers should be kept in an air-tight box with a shallow jar at the bottom for water as needed. In damp weather remove the water, in dry weather let the water remain. Rollers when out of use any length of time should be put away with the ink on them to protect their surface from the action of the atmosphere. It would be difficult to find a pressman who could be induced to believe there was anything for him to learn in the making of rollers. Several things enter into the choice of composition, such as quality of ink used, climate, class of work, requirements of presses, etc. The cores should be cleansed by scraping, or, if of wood, by scalding in strong lye or soapsuds, then dried. New rollers should be washed in sperm or coal oil before used. It will prevent the strong suction. Turpentine is better than benzine for removing coloured inks. Never use lye on new rollers.

Wear and Tear of Machine Rollers.—Someone asks : "Can you give me a remedy for preventing machine distributing rollers from wearing at the ends ? We run our weekly edition off on a quad-demy Wharfedale machine at the rate of 1,100 per hour ; and before we have printed six or seven hundred the rollers begin to wear. I may here state that the inking rollers run for months without renewing." —There may be more than one reason why the distributors wear at the ends. Are the stocks clean and free from grease ? A good plan to help composition to stick well to the stock is to tie round the latter some tape or band previous to casting. Perhaps the rollers are not sufficiently supported in their runners ; distributors should be so fixed that they barely touch the inking-table, and should not be allowed to be so low in the runners as to cause them to bump against the ink-table every time it returns. If too low, the runners should be packed, so as to lift the rollers to barely leave their own weight on the ink-table. A good way to prevent rollers pulling is to cut off the end of the composition slopingly, and hold the end of the roller in a gas flame : this will slightly melt the composition, which may be drawn well over the stock with the finger, effectually making the end air and water tight. If the composition at the end of the roller is not fast to the stock when washed, water gets under, with the consequence that the composition swells, making the roller thicker at the end than in the centre.

Renovating Old Rollers.—When rollers have been lying for weeks with a coating of ink dried on to the surface—a circumstance that often occurs, more especially when coloured inks have been used—get an ordinary red paving brick (an old one with the edges worn away will be the best), place the roller on a board, then dip the brick in a trough of cold water and work it gently to and fro on the surface from end to end, taking care to apply plenty of water, dipping the brick in repeatedly; and in a short time the ink will disappear. Nor is this all : for if a little care and patience is exercised, it will put a new face to the roller, making it almost equal to new ; the coating of ink having, by keeping the air from the surface, tended to preserve the roller from perishing. Sponge off clean.

To Cast Rollers.—See that the roller-mould is perfectly clean ; make a mop and with it oil carefully every part of the interior of the mould. Now turn your attention to the stock ; be quite sure that it is perfectly dry, and if the composition is apt to slip off it, bind some string very lightly upon it, then place it in the mould, being very careful that it stands true in the middle ; fit on the guide at the top, and fasten the stock down to the mould with string lest it should rise ; then warm the mould all over. Meanwhile your composition will have been melting : take care that it does not boil, and stir it about occasionally. Never re-melt old composition without a good proportion of new, and if the old is very hard, you may add some treacle. It is best to use one of the special kettles sold by the printers' furnishers, as otherwise there may be some difficulty in getting the composition to melt properly. When it is quite melted, carefully pour it into the mould, filling it to the level of an inch or two higher than the end of the stock to allow for shrinkage. Let it stand for about twelve hours; then prepare to draw the roller out of the mould. If it will not come readily, one man must hold the stock and another the mould, and they must pull without jerking in opposite directions. Pushing upon the lower end of the roller may perhaps be necessary also, but if the mould has been properly oiled there ought not to be much difficulty in drawing it. When out, trim the ends, and hang it up in a dry, cool place for a day or two before using.

Another Recipe for Casting Rollers.—Messrs.

Fairholme and Co. have just issued instructions as to the treatment of their composition, and they will apply to composition generally. They are as follows :—Cut the composition into small pieces to assist it melting. If too firm, add when melted a sufficient quantity of either glycerine or molasses. Water must not be added, as the composition will shrink on the rollers after casting if this is done. Have the mould well warmed before filling, which is best done by plunging the mould into hot water for a few seconds. This prevents the composition getting chilled ere it reaches the bottom. Carefully oil the inside of the mould. *Renewing.—* Sponge the face of the roller with warm water ; scrape off the face thoroughly with a knife previous to removing the composition from the stock, and cut it up small. If the roller has been used only a short time it may be melted as readily as new composition. If too firm, add glycerine or molasses, as mentioned above. It sometimes happens that small pieces refuse to melt readily along with the bulk of the composition. Such refractory pieces always float on the top. On no account persevere in trying to melt these pieces, but remove any such at once from the pot, and cast your roller, as by continuing the heat there is great danger of overboiling the whole, when the composition becomes leathery and ceases to

H

be a liquid. In remelting, a certain percentage of refractory pieces is sure to occur, and must be sacrificed. Wash the stocks after removing the old composition, to remove all traces of grease. On no account allow the composition to remain in the pot after it is melted, but pour immediately into moulds. Every minute that the melted composition remains it is deteriorating. Crystallization is caused by boiling, and the composition becomes stiff and stringy.

Good Wearing Rollers.—It is stated that rollers made from Chinese sugar-cane molasses are far superior to those made with any other kind. The syrup will bear long boiling without granulation, and when cast into a roller is much tougher, more elastic, and has better suction than those made by the material in common use.

Preserving Rollers.—A simple process for preserving and renovating ink rollers, and adding greatly to their longevity, is as follows :—A steam jacket is added to the roller closet, and numerous fine jets are so arranged as to play gently upon the rollers within. These jets thoroughly cleanse the surface of the roller, the skin on its face disappears, the body of the roller absorbs a portion of the heated vapour, and the whole is kept in a fresh, elastic condition, ready for work without further preparation. Experiments by practical men seem to show that the contrivance possesses considerable value, and is likely to prove very economical in large printing establishments.

The Treatment of Rollers.—Rollers, like children, who are subject to so many ailments, must be 'well attended in order to be thoroughly healthy. There are times when they are affected by the heat, cold, damp, and dry, and must be treated accordingly. One of the secrets of good printing is in using good rollers. A good roller should be moderately soft and elastic, and not be too new. It depends upon the manufacture how old it should be before being fit to work : some rollers may be used a few hours after coming out of the mould, while others of a damper nature should not be used for a week. A new roller should have a moderately dry face before putting it on the machine, otherwise it will not take the ink properly, and will friar the type, and the composition is likely to draw off the stock. They should be washed with turps, and kept in a thoroughly dry place, slightly oiled. Old rollers should be well washed with lye and rinsed with water; they should be smothered over with thin broadside ink, but should occasionally be scraped and re-inked to prevent drying on. In hot weather rollers should be kept in a cool place. Wavers tearing away at ends is frequently the case, especially in fast speed machines; to prevent it a little oil is often put on ends of slab, but this is a dangerous plan, as it works in and causes the work to run pale. Better methods are—Cut away the composition, tapering within a few inches of ends; or get a bucket of hot water and dip the ends in until they become taper; or screw a piece of wood on to the end of ink-slab, about two inches wide in centre, tapering off to a quarter of an inch at ends. This will cause the wavers to start revolving from centre instead of at ends, and will thus preserve them.

To Keep Rollers.—In Germany the following preservative of rollers when not in use is often applied :—Corrosive sublimate, 1 drachm ; fine table salt, 2 oz.; put together in half a gallon of soft water. It is allowed to stand twenty-four hours, and is to be well shaken before using. Sponge the rollers with the mixture after washing.

Lubricating Roller Moulds.—Sperm and lard oils are the best. If they are properly used, no trouble will be experienced in drawing the rollers.

To Clean Rollers used for Printing Copying Inks.—It is best to avoid water, which always weakens them. Spirits of wine proves much more efficient ; it takes the ink off immediately, does not injure the rollers, and, as it vaporizes almost immediately, they may be used directly.

Rollers should be cleaned at least once a day, never mind what is said to the contrary. A roller, like our faces, wants washing occasionally.

Lithographic Rollers.—Take care of your rollers. A good roller is worth at least as much as a good printer. A roller that is not to be used in three days should be left in medium varnish. A roller not to be used for eight days should be covered with tallow. A roller not to be used for months should be rolled up in glycerine. Provers should keep at least six rollers—one for black, one for red, one for yellow, one for blue, one for tint, and one good roller with sharp grain for crayon drawings. The blue roller can be used for green and purple also; this will not last long, because blue, green, and purple inks eat into the roller, and the frequent washings necessary will cause it to wear out soon. Scrape the roller alternately with and against the grain—that is, first in one direction and then in the opposite. A roller too soft fills up the work; one too hard takes the work off the stone. A roller too smooth slips over the stone, and takes the work off also. A roller with fine grain makes a solid impression, but rusy. A roller with a rough grain makes a sharp impression, but not a solid one. An uneven roller should not be used at all. A bad roller is a detriment to every printer using it, for the best of them can do nothing with it. The best roller, even if costing double in price, is cheapest.

INKS, VARNISHES, ETC.

HINTS on Colours.—It is not desirable to rely upon make-shifts, but there are occasions when printers are from circumstances obliged to decline work, not from a want of knowing how to produce it, but from a lack of material to do it with. We all know how small quantities of coloured ink dry in cans. Provide yourself with a bottle of inkoleum, which will reduce and refine the oldest and driest coloured inks to a consistency that will make them work free without injury to the colour.

RED INK.

If for fine work, procure a few ounces of the best vermilion powder, and mix it well with thick varnish, allowing the varnish to carry all the colour it possibly can ; when you commence grinding with the colour stone and muller, do not despair if you find the mixture form itself into a ball resembling a lump of india-rubber ; if this should follow, add plenty of muscle to the muller, and it will assume the consistency required. If, when you attempt to work it, you find the colour adhere to the inking-table, and the varnish attaches itself to the roller and the printing looks poor and without body, add a little of the inkoleum. If you require the red ink for surface printing, use the middle varnish, and for poster or show work use thin varnish by itself.

DARK BLUE INK.

This may be produced in the same way, by purchasing a few ounces of dark Chinese blue, but requires a hundred per cent. more grinding than the vermilion, and must be ground in very small quantities at a time ; by simply spreading with a palette knife you will see if it has a rough or smooth appearance—it will recover its smoothness instantly ; do not be deceived by it without this test.

LIGHT BLUE INK.

A very pretty blue may be made by substituting a few ounces of Antwerp blue ; proceed as in the former directions, but it will take more grinding than the vermilion, but less than for the Chinese.

LIGHT GREEN INK.

To make this very little grinding is required, if the colour obtained is of the best quality.   In other respects proceed as with the vermilion.

DARK GREEN INK.

Add to the above a small quantity of dark Chinese blue.

CRIMSON INK.

Use the purest carmine, well ground to a thick consistency; a very small portion of this added to the vermilion will give it " fire," and kill the brick-dust appearance.

PINK INK.

Lower the carmine with pure flake white.

COMMON BROWN INK.

Add a small quantity of black to the vermilion.

TINTS, ETC.

Grey sets off a colour better than either black or white. White, gold or black will serve as an edging to any colour. A white ground has a tendency to make colours upon it appear darker, while a black ground has a contrary effect.   In the association of two tones of one colour, the effect will be to lighten the light shade and darken the other.   The fact that incongruous colours are often harmoniously combined in nature is no guarantee that they may be similarly applied in art.

Type-Writer Ribbon Ink.—Aniline black or violet, half ounce ; pure alcohol, 15 oz. ; concentrated glycerine, 15 oz. Dissolve the aniline in the alcohol and add the glycerine.

A Good Dryer for Poster Inks.—Spirits of turpentine, one quart ; balsam copaiba, six ounces. Add a sufficient quantity of ink to thin it to a proper consistency for working. This compound is one of the best that can be used as a dryer ; it brightens the inks and makes them work freely. Ruling-inks can be made to dry quickly by adding half a gill of methylated spirits to every pint of ink. The spirit is partly soaked into the paper and partly evaporates, and it also makes the lines firm.

A Good Black Varnish.—By means of a gentle heat dissolve 50 parts of powdered copal in 400 parts of the oil of lavender ; add 5 parts of lampblack and 1 part of powdered indigo.

To Make Stencil Ink.—A good and cheap stencil ink, in cakes, is said to be obtained by mixing lampblack with fine clay, a little gum arabic or dextrine, and enough water to bring the whole to a satisfactory consistency.

A Bronze or Changeable Hue.—A bronze or changeable hue may be given to inks with the following mixture : Gum shellac, $1\frac{1}{2}$ lb., dissolved in one gallon of 95 per cent. alcohol or Cologne spirits for twenty-four hours. Then add 14 oz. aniline red. Let it stand for a few hours longer, when it will be ready for use. When added to good blue, black, or other dark inks, it gives them a rich hue. The quantity used must be very carefully apportioned. In mixing the materials, add the dark colour sparingly at first, for it is easier to add more, if necessary, than to take away, as in making a dark colour lighter you increase its bulk considerably.

Printing Oil.—Linseed oil, 1 quart ; 1 pint of rape oil ; 1 oz. balsam of copaiba ; $\frac{1}{2}$ oz. of pitch ; $\frac{1}{2}$ oz. of amber oil ; $\frac{1}{2}$ oz. of white lead.

Handling of Small Cans of Ink.—A good plan is to remove the lid and place the tin in a black jar or pot slightly larger. Throw away the lid of the can and keep the ink workable by pouring on the top a little glycerine. The lid of the jar is then put on, and the whole can be used without defiling the fingers.

Reducing Tint Inks.—Colour for tint work is invariably toned down with varnish. If the colour appears mottled, add a little flake white to give it body.

Green Transparent Varnishes for Metals.—Grind a small quantity of Chinese blue with double the quantity of finely powdered chromate of potash (it requires the most perfect grinding); add a sufficient quantity of copal varnish thinned with turpentine. The tone may be altered by adding more or less of one or the other ingredients.

Stamping Inks.—*Red*: dissolve one-fourth of an ounce of carmine in two ounces of strong ammonia, and add one drachm of glycerine and three-fourths of an ounce of dextrine. *Blue*: rub one ounce of Prussian blue with enough water to make a perfectly smooth paste; then add one ounce of dextrine, incorporate it well, and finally add sufficient water to bring it to the proper consistency.

Improved Dryer for Printing Ink.—Many of the "dryers" that are now sold are very unsatisfactory. Printed work for which they are used not unfrequently sets off, and is at all times difficult to handle. If the greatest care is not used there is "spoilage." Here is a recipe supplied by a practical chemist, which deserves to be tried:—A small quantity of perfectly dry acetate of lead or borate of manganese in impalpable powder will hasten the drying of the ink. It is essential that it should be thoroughly incorporated with the ink by trituration in a mortar.

Component Parts of Printing Inks.—Varnish composed of linseed oil, resin and soap, and the pigment; in the case of black ink, lampblack and indigo with Prussian blue.

Method for Making Red Ink.—Red printing ink may be made in this way:—Boil linseed oil until smoke is given off. Set the oil then on fire, and let it burn until it can be drawn out into strings half-an-inch long. Add one pound of resin for each quart of oil, and one half-pound of dry, brown soap cut into slices. The soap must be put in cautiously, as the water in the soap causes a violent commotion. Lastly, the oil is ground with a pigment on a stone by means of a muller. Vermilion, red lead, carmine, Indian red, Venetian red, and the lakes are all suitable for printing inks.

Japanners' Gold Size.—Powder finely of asphaltum, litharge, or red lead, each one ounce ; stir them into a pint of linseed oil, and simmer the mixture over a gentle fire or on a sand-bath till solution has taken place, scum ceases to rise, and the fluid thickens on cooling. If too thick when cool, thin with a little turpentine.

Copying Ink.—An ink to be used without press or water can be made by taking three pints of jet-black writing ink and one pint of glycerine. This, if used on glazed paper, will not dry for hours, and will yield one or two fair, neat, dry copies, by simple pressure of the hand, in any good letter copy-book. The writing should not be excessively fine, nor the strokes uneven or heavy. To prevent " setting off," the leaves should be removed by blotting-paper.

Preserving Ink.—Coloured inks can be kept from " skinning " by pouring a little glycerine, oil, or water on the top, and closing the can tightly.

Endorsing Ink.—Is made of Prussian blue well ground up with linseed oil.

To Make White Ink.—Mix pure, freshly precipitated barium sulphate or flake white with water containing enough gum arabic to prevent the immediate settling of the substance. Starch or magnesium carbonate may be used in a similar way. These must be reduced to impalpable powder.

Gold Ink.—For making gold ink take equal parts of iodide of potassium and acetate of lead ; put them on a filter, and pour over them twenty times the quantity of warm distilled water. As the filtrate cools iodide of lead separates in golden scales. This is collected when the filtrate has quite cooled, washed with cold water on a filter, and rubbed up for an ink with a little mucilage. The ink must be shaken every time it is used.

Hardening Gloss for Inks.—Dissolve gum arabic in alcohol or a weak solution of oxalic acid, and add it in small quantities to the ink in use.

Turkey Red by a New Process.—An interesting method has been lately introduced for producing the beautiful Turkey red from alizarine. A certain quantity of Turkish red oil is dissolved in water, and a certain percentage of alizarine added, also tannin. This mixture is slowly heated to a boiling temperature, and a solution of aluminum sulphite added, of 1.1014 specific gravity, which has been previously mixed with twenty-two per cent. of soda crystals. On prolonged boiling, the alizarine lake separates out, this being freed from excess of oil by washing with ether. It then forms a powder of splendid carmine-red colour, which is constant in the light, and is not attacked by dilute acids and alkalies. It still contains a certain quantity of oil, which cannot be removed by ether, but which gives the lustre to the preparation. When mixed in a thorough manner with water, the lake can be used for dyeing tissues in shades similar to those produced by eosine.

Stencil Blue Ink.—Shellac, 2 ounces; borax, 2 ounces ; water, 25 ounces; gum arabic, 2 ounces ; and ultra-marine sufficient. Boil the borax and shellac in some of the water till they are dissolved, and withdraw from the fire. When the solution has become cold, add the rest of the 25 ounces of water and the ultra-marine. When it is to be used with the stencil it must be made thicker than when it is to be applied with a marking brush.

Printing Ink.—The following hint is valuable, for it is only, as a rule, known to old-fashioned pressmen. The scrapings and dregs of ink-casks, skins of ink, scrapings of rollers, and waste ink generally, if sent to a small ink-maker to be remade up with some fresh ink, will turn out the best ink for a printer's use that can be obtained. We have known splendid specimen-work done with ink thus compounded, some of the material having been laid aside for years as waste. If the printer has an ink-grinding machine he may make the ink himself.

To Make Chinese White.—Take as much as required of zinc white finely ground, put it on a marble or glass slab, mix it into a cream of the required consistence by adding mucilage of gum tragacanth, grinding with a glass muller. For quantity required to fill an ordinary sized Chinese white bottle, add to the above 10 or 12 drops of thick gum arabic and 5 or 6 drops of pure glycerine, grinding well together.

Bookbinders' Varnish.—The best receipt for making varnish for full calf extra work is as follows :—3 pints of spirits of wine of 40 per cent., 8 oz. shellac, 8 oz. sandarach, 2 oz. mastic in drops, 2 oz. Venice turpentine ; apply lightly on the book with a piece of cotton wool, a small sponge, or a brush.

How to Improve Black Ink on Creamy Paper. —The brownish hue seen in black ink when printed on toned paper may be obviated by mixing a little blue ink with the black.

Varnish in Colour Printing.—In the manipulation of colours nothing forms a more important factor than the varnishes with which the pigments are mixed. It matters not how pure and well-ground these may be if the consistency of the varnish is unsuited to the peculiar colour to be made into printing ink. A practical knowledge of this branch of work is seldom attained ; although we know no reason why any pressman neglects to learn all that pertains to colours and their treatment in the manufacture of ink, except it be foolish negligence. To be a good pressman a wide range of study in all that pertains to the manipulation of colours is necessary. It matters not how thorough a person's ability may be to make ready a forme, his dexterity in getting it worked off, or his precision in the management of work; if he lacks sound, practical knowledge of colour-making, he is still deficient in his business. To such a one colours are a labyrinth of confusion, and a lack of this knowledge is at times the cause of large outlay and waste to employers, who, from a want of personal practical insight, allow their workmen to guess what is required. Nor should there be a want of attention or information in the matter of paper used in printing. This part of a printer's study is as important as colour manipulation, for there are excellent inks that will not work on some papers. To overcome this a familiarity with paper surfaces and the materials used in their manufacture is an essential matter.

Dryer for Ruling Inks.—Ruling inks are made to dry quickly by using half a gill of methylated spirits to every pint of ink. The spirit is partly soaked into the paper and partly evaporates; it also makes the lines firm.

Imperishable Ink.—The nearest approach yet discovered is made by grinding up good Chinese or Japanese Indian ink in a saturated solution of borax. The Indian ink is itself imperishable, but can be washed off from paper before the borax solution fixes it.

Imitation Gold Lacquer.—The preparation of the real Chinese gold lacquer is a secret, but an excellent imitation may be prepared by melting two parts of copal and one of shellac, so as to form a perfectly fluid mixture, and then adding two parts of hot boiled oil. The vessel is then to be removed from the fire and ten parts of oil of turpentine gradually added. To improve the colour addition is made of a solution in turpentine of gum gutta for yellow and dragon's blood for red in sufficient quantity to give the desired shade. The Chinese apparently use tinfoil to form a ground, upon which the lacquer varnish is laid.

To Make Lye.—Use one pound of pearlash to three quarts of water, or one pound of potash to five quarts of water.

Indelible Ink.—A cheap indelible ink can be made thus : Dissolve in boiling water twenty parts of potassa, ten parts of finely-cut leather chips, and five parts of flowers of sulphur, the whole heated in an iron kettle until it is evaporated to dryness. Then the heat is continued until the mass becomes soft, great care being taken that it does not ignite. The pot is now removed from the fire, allowed to cool, water is added, and the solution strained and preserved in bottles. This ink will readily flow from the pen.

Restoring Writing on Old Deeds.—The faded ink on old parchments may be so restored as to render the writing perfectly legible. Moisten the parchment or paper with water, and then pass over the lines in writing a brush which has been dipped in a solution of sulphide of ammonia. The writing will immediately appear dark, and this colour it will preserve. Parchment records treated in this way in the Museum at Nuremberg ten years ago are still perfectly legible. On paper, however, the colour gradually fades away; but by the application of the sulphide it can be restored; for the iron which enters into the composition of the ink is transformed by the reaction into black sulphide.

Luminous Paint.—White luminous paint has been known for some years, and chiefly used for watch dials, match boxes, and other small objects, but all attempts to produce luminous paints of various colours have hitherto failed, owing to the fact that the sulphuret of calcium, the principal constituent, is decomposed by the metals contained in the usual colours, whereby not only the luminosity, but also the colour, is destroyed. Lately, however, G. Schatte, of Dresden, has succeeded in producing red, blue, green, and otherwise coloured paints, which have the quality of shining in the dark with the same colour which they possess in the daylight. It is easy to see that a variety of artistic effects may be produced by means of such business colours.

Black Bordering.—The best shining black ink, used for mourning paper, which has up to the present time been kept a secret by makers, may be prepared of lampblack, borax, and shellac. The ink is made as follows:—In 1 litre of hot water 60 grammes of borax are dissolved, and to this solution three times the quantity of shellac is added. After this mixture has been properly dissolved, the necessary quantity of lampblack is added, the whole being constantly stirred. Should the lustre not be satisfactory, more shellac is added.

How to Remove Aniline Ink from the Hands.—Aniline inks are now in common use, especially in connection with the various gelatine tablets for multiplying copies of written matter. Upon the hands it makes annoying stains, difficult of removal by water or acids. They may be easily washed out by using a mixture of alcohol, three parts, and glycerine, one part.

Composition for Bronze Work is a mixture of chrome yellow and varnish. The chrome is well ground with a muller into the varnish. This gives the bronze, especially gold, a fuller tint than if the plain varnish only is used. It answers equally well for copper, citron, or emerald bronze. To give silver bronze a deep appearance ordinary dark blue ink may be used.

Picture Varnish.—Gum mastic, 6 ounces; pure turpentine, 4 drachms; camphor, 2 drachms; oil of turpentine, 19 ounces. Add first the camphor to the turpentine, and heat them over a water-bath until solution is effected. Then add the gum mastic and the essential oil of turpentine, and finely filter through cotton wadding. The varnish, on being kept several months, improves in toughness and brilliancy. It is to be applied with a fine varnish brush; when quite dry it will stand washing without injury.

Pheasant Colour.—Whiting, 9 parts; 9 parts flint; 1 part prepared oxide cobalt.

Rubber Stamp Ink.—The following proportions are said to give an excellent ink, which, while not drying up on the pad, will yet not readily smear when impressed upon the paper:—Aniline red (violet), 90 grains ; boiling distilled water, 1 oz.; glycerine, $\frac{1}{2}$ teaspoonful ; treacle, half as much as glycerine. The crystals of the violet dye to be powdered and rubbed up with the boiling water, and the other ingredients stirred in.

Another Method.—It is said another endorsing ink, which does not dry quickly on the pad, and is quickly taken by the paper, can be obtained by the following receipt :—Aniline colour in solid form (blue, red, &c.), 16 parts, 80 parts boiling distilled water, 7 parts glycerine, and 3 parts syrup. The colour is dissolved in hot water, and the other ingredients are added whilst agitating. This endorsing ink is said to obtain its good quality by the addition of the syrup.

Quick Drying Indian Ink.—This ink for drawing may be improved so that even the thickest lines will quickly dry by adding one part of carbolic acid to eighty of the Indian ink. If too much is added it may be rectified by putting in more Indian ink. If the mixture is properly formed the ink is as easy to draw with as it is without carbolic acid, but dries quickly, and may even be varnished without discharging.

Quick Drying Varnish for Paper Book Covers.—Add to the varnish a solution made as follows:—6 ounces mastic, in drops ; 3 ounces coarsely powdered glass, separated from the dust by a sieve ; 32 ounces spirits of wine of 49°. Place the ingredients in a sand-bath over a fire, and let them boil, stirring well. When thoroughly mixed introduce 3 ounces spirits of turpentine, boil for half an hour, remove from the fire, cool, and strain through cotton cloth. Great care in manipulation is requisite to avoid a conflagration. Use a close fire and watch incessantly.

Prepared Blue for Printing.—Oxide cobalt, 10 lbs. ; 12 oz. red lead. The above to be calcined in oven. Then pound, add 10 lbs. of whiting, and send to mill to be ground.

Sympathetic Ink.—An ordinary solution of gum camphor in whisky is said to be a permanent and excellent sympathetic ink. The writing must be done very rapidly, as the first letters of a word have disappeared by the time the last are written. Dipping the paper in water brings it out distinctly, and it becomes invisible again when the paper is dried. It can be brought out repeatedly without affecting its vividness.

Cheap Black Ink.—A German paper gives the following :—Extract of Campeachy wood 100 parts, lime water 800 parts, phenol (carbolic acid) 3 parts, hydrochloric acid 25 parts, gum arabic 30 parts, red chromate of potash 3 parts. The extract is first dissolved in the lime water on a steam-bath with frequent stirring or shaking; after this the carbolic and hydrochloric acids are added, which change the red colour to a brownish yellow. It is then heated half an hour on steam-bath and set aside to cool. It is next filtered, and the gum and bichromate, dissolved in water, are added. Enough water is added to make up the solution to 1,800 parts. This ink, red when first used, quickly turns black.

Erasible Ink Stencil can be made of lampblack or bone-black, one ounce; yellow soap, one drachm; water sufficient; mix and beat into a paste.

To Test Japan and Varnish.—Japan, like varnish, must be good to give entire satisfaction, and much damage is done by using a poor article. One way of testing a japan is to spread some on a piece of glass and leave it in the direct rays of the sun. When it has entirely lost its fluidity, scratch it lightly with the nail, and if it falls in powder without cracks, its quality is proven good. This is also said to be a good way of testing varnish. The liquid which begins to enamel in places is of an inferior quality.

Violet Aniline Ink for 'Graph Printing can be made by dissolving one part nigrosine in five parts of water and one part of alcohol, adding afterwards one part of glycerine. Violet is used because it gives more copies than any other colour.

Indelible India Ink.—Draughtsmen are well aware of the fact that lines drawn on paper with good India ink which has been well prepared cannot be washed out by mere sponging or washing with a brush. Now, however, it is proposed to take advantage of the fact that glue or gelatine, when mixed with bichromate of potassa and exposed to the light, becomes insoluble, and thus renders India ink, which always contains a little gelatine, indelible. Reisenbichler, the discoverer, calls this kind of ink "Harttusch," or hard India ink. It is made by adding to the common article, when making, about one per cent. of bichromate of potash in a very fine powder. This must be mixed with the ink in a dry state, otherwise, it is said, the ink could not be ground up easily in water. Those who cannot provide themselves with ink prepared as above in the cake, can use a dilute solution of bichromate of potash in rubbing up the ink; it answers the same purpose, though the ink should be used thick, so that the yellow salt will not spread.

Ticket Writers' Ink is made of good black ink with liquid gum added in the proportion of half an ounce of gum to a gill of ink. Set in a warm place and shake up occasionally till well mixed. If wanted very glossy add more gum, always remembering that increasing the quantity of gum makes the ink less easy in working.

An Elastic Flexible Varnish for Paper, which may be applied without previously sizing the article, may be prepared as follows:—Crush transparent and clear pieces of dammar into small grains; introduce a convenient quantity—say forty grains—into a flask, pour on it about six ounces of acetone, and expose the whole to a moderate temperature for about two weeks, frequently shaking. At the end of this time pour off the clear saturated solution of dammar in acetone, and add, to every four parts of varnish, three parts of rather dense collodion; the two solutions are mixed by agitation, the resulting liquid allowed to settle, and preserved in well-closed phials. The varnish is applied by means of a soft beaver-hair pencil, in vertical lines. At the first application it will appear as if the surface of the paper were covered with a thin

I

white skin. As soon, however, as the varnish has become dry, it presents a clear, shining surface. It should be applied in two or three layers. This varnish retains its gloss under all conditions of weather, and remains elastic ; the latter quality adapts it especially to topographical crayon drawings and maps, as well as to photographs.

Marking-Ink Without Nitrate of Silver.—One drachm of aniline black is rubbed up with 60 drops of strong hydrochloric acid and $1\frac{1}{2}$ oz. of alcohol. The resulting liquid is then to be diluted with a hot solution of $1\frac{1}{2}$ drachms of gum arabic in 6 oz. of water. This ink does not corrode steel pens, and is unaffected either by concentrated mineral acids or by strong lye. If the aniline black solution is diluted with a solution of $1\frac{1}{2}$ oz. of shellac in 6 oz. of alcohol, instead of with gum water, an ink is obtained which, when applied to wood, brass, or leather, is remarkable for its black colour.

Chrome Writing Ink.—This useful, cheap, and almost unalterable ink can be made according to the following recipe : Distilled water, 1,000 parts (by weight); logwood extract, 16 parts ; carbonate of soda (cryst.), 2 parts ; chromate of potassium, 1 part. Dissolve the logwood extract in 900 parts of water by aid of heat, and let it stand to settle ; draw off the clear liquid, heat to boiling, and add the carbonate of soda ; lastly add, drop by drop, with constant stirring, the chromate (yellow chromate) previously dissolved in 100 parts of water. The colour is not fully developed at once, but on standing for a few hours gradually deepens to a full bluish black. The ink thus prepared flows well and dries quickly. The addition of a trace of clove oil will prevent mouldiness.

Indestructible Ink.—India ink, and many other varieties of ink containing considerable quantities of carbon, are practically indestructible—that is to say, they are far more permanent than the material on which they are used, as paper in time becomes exceedingly brittle and friable. A very good formula is the following, which is said to yield an ink very much resembling that forming the characters upon the

Egyptian papyrus :—Make a solution of gum lac in an aqueous solution of borax, and add to this a sufficient quantity of lamp-black to give the proper black coloration.   This ink is claimed to be almost indestructible, resisting both time and chemical agents, and yielding a beautiful lustrous black.   The printing press has, for the general preservation of literary treasures, largely reduced the necessity for indestructible materials upon which to record them permanently, since they may readily be duplicated and distributed.   But the question of the preservation of important public records upon more permanent materials than wood paper and indifferent inks, will doubtless make itself felt as a grave necessity before many more years have passed.

Frost-proof Ink.—Aniline black one drachm, rub with a mixture of concentrated hydrochloric acid one drachm, pure alcohol 10 ounces.   The deep blue solution obtained is diluted with a hot solution of concentrated glycerine $1\frac{1}{2}$ drachms, in four ounces of water.   This ink does not injure steel pens, is unaffected by concentrated mineral acids or strong alkalies, and will not freeze at a temperature of 22 or 24 degrees below zero.

Ink for Writing on Glass.—This can be made by dissolving in strong hydrofluoric acid enough gum arabic to make the liquid flow readily from a pen without being viscid, and to colour it with cudbear or some other colouring matter which will stand an acid, so that the writing may be visible to the eye.   The solution must *not* be made in a glass vessel, but is best made in a rubber bottle.   When a portion is poured out, it should be poured into a platinum or lead vessel, or into a rubber nipple set up in a small wide-mouthed bottle.   Care must be taken that the acid be not brought in contact with the skin, as it produces painful and troublesome sores.

To Give a Silky Effect to Ink.—The new style of silk and satin lettering for showcards is produced as follows : Take an impression in any colour mixed with varnish, and dust it as in bronzing with powdered asbestos.

Cardinal Ink for Draughtsmen.—The solution of carmine lake in caustic aqua ammonia is attended with this disadvantage—that, in consequence of the alkaline properties of ammonia, the cochineal pigment will in time form a basic compound, which, in contact with a steel pen, no longer produces the intense red, but rather a blackish colour.   To avoid this, prepare the ink as follows :—Triturate 10 grains of pure carmine with 150 grains of acetate of ammonia solution and an equal quantity of distilled water, in a porcelain mortar, and allow the whole to stand for some time. In this way a portion of the alumina, which is combined with the carmine dye, is taken up by the acetic acid of the ammonia salt, and separates as precipitate, while the pure pigment of the cochineal remains dissolved in the half-saturated ammonia. It is now filtered, and a few drops of pure white sugar syrup added to thicken it.   In this way an excellent red drawing ink is obtained, which holds its colour a long time. A solution of gum arabic cannot be employed to thicken this ink, as it still contains some acetic acid, which would coagulate the bassorine, one of the natural constituents of gum arabic.

Old Manuscript Ink.—The following formula is said to have been in use in 1654, and to have produced an ink of great permanency, if one may judge from manuscripts written by the person who is the authority for the formula :—One and one-half drachms of coarsely powdered galls, six drachms of sulphate of iron, ten drachms of gum arabic, and one pint of soft water, are to be placed in a bottle, which is to be securely stoppered and placed in the light (sunlight if possible).   Stir occasionally until the gum and copperas are dissolved, after which the bottle should be shaken daily.   In the course of four to six weeks the ink will be fit for use. The addition of ten drops of carbolic acid will prevent the formation of mould.

Diamond Ink for Glass.—The ink used for writing or etching upon glass is composed of ammonium fluoride dissolved in water and mixed with three times its weight of barium sulphate.

Lithographic Ink.—The following is a recipe likely to be useful to amateur lithographers and others:—Wax, 16 parts; tallow, 6 parts; hard tallow soap, 6 parts; shellac, 12 parts; mastic in tears, 8 parts; Venice turpentine, 1 part; lampblack, 4 parts. The mastic and lac, previously ground together, are to be heated with care in the turpentine; the wax and tallow are to be added after they are taken off the fire, and when the solution is effected the soap shavings are to be thrown in. Lastly, the lampblack is to be well intermixed. Whenever the union is accomplished by heat, the operation is finished; the liquor is left to cool a little, then poured out on tables, and, when cold, cut into square rods. Lithographic ink of good quality ought to be susceptible of forming an emulsion so attenuated that it may appear to be dissolved when rubbed upon a hard body in distilled or river water. It should be flowing in the pen, not spreading on the stone; capable of forming delicate traces, and very black to show its delineations. The most essential quality of the ink is to sink well into the stone, so as to reproduce the most delicate outlines of the drawing, and to afford a great many impressions. It must, therefore, be able to resist the acid with which the stone is moistened in the preparation, without letting any of its greasy matter escape.

White Ink for Pen Drawing.—Mix pure freshly precipitated barium sulphate or flake white with water containing enough gum arabic to prevent the immediate settling of the substance. Starch or magnesium carbonate may be used in a similar way. They must be reduced to impalpable powders.

Vanishing Black Writing Ink.—To make an ink, black at the time of writing, but which shall disappear after a short time, boil nut-galls in aqua vitæ, put Roman vitriol and sal ammoniac to it, and when cold dissolve a little gum in it. Writing done with this ink will vanish in twenty-four hours.

Printers' Size.—Common soda, ¼ lb.; 1 lb. soft soap; 1 gallon water.

A Recipe for Marking Ink.—Take nitrate of silver 11 grains, dissolve in 30 grains of aqua ammonia. Dissolve 20 grains of gum arabic in 85 grains (2½ teaspoonfuls) of rainwater. When the gum is dissolved, put in the same phial also 92 grains of carbonate of soda. When all are dissolved, mix the contents of both phials together, and place the phial containing the mixture in a basin of water, and boil several minutes, or until a black compound is the result. When cold it is ready for use.

New Invisible Ink.—To make the writing or drawing appear which has been made upon the paper with the ink, it is sufficient to dip it into water. On drying, the traces disappear again, and reappear by each succeeding immersion. The ink is made by intimately mixing linseed oil, 1 part; water of ammonia, 20 parts ; water, 100 parts. The mixture must be agitated each time before the pen is dipped into it, as a little of the oil may separate and float on top, which would, of course, leave an oily stain upon the paper.

A New Varnish.—The base consists of paper treated with nitro-sulphuric acid and camphor dissolved in alcohol. The varnish is composed of this ingredient, with acetic ether, sulphuric ether, castor oil, Venetian turpentine, methylated alcohol, acetate of amyl, and pure crystallizable acetic acid in definite proportions. This varnish is said to have the following properties :—It is unaffected by water and humidity, and will also resist weak and concentrated acids, and likewise alkalies if contact be not too prolonged ; it takes a fine polish. It can only be applied to surfaces warmed either by the sun, before the fire, or in an oven at the temperature of 35° to 45° centigrade. If applied to a cold surface it becomes white, and the coat is wanting in coherence and possesses none of the qualities of the same varnish applied to a warm surface. Any kind of brush may be used for applying it. It dries in two minutes at the outside, so that twenty-five or thirty coats may be applied per hour. It may be rendered more flexible, and may then be applied to cold surfaces, by adding crystallizable acetic acid and acetate of amyl.

Black Enamel for Brass Plates.—The enamel on brass signs, by which the letters cut into the metal are filled up, is made by mixing asphaltum, brown japan, and lamp-black into a putty-like mass; fill in the spaces and clean the edges with turpentine.

Indelible Aniline Ink.—It is necessary, to render this ink indelible on paper, to coat the reproduction with some preparation. An excellent compound consists of collodion dissolved to the consistency used by photographers, with two per cent. stearine added.

Ink for Enamelled Cards.—For printing on enamelled cards add to good quality printing ink a composition of copal varnish 1 oz., mastic varnish 1 oz. Mix together, add twenty drops to the ink, and print. Old enamelled cards are better to print upon than new ones.

Calico Printing Dye.—The black dye used is obtained from myrabolans, a highly astringent fruit imported from India and other eastern countries.

Recipe for Printing Gloss Ink.—Take one pound of sugar of lead, one pound of white copperas, and half a pound of litharge oil, in a jar, which should be kept in a cool place. This preparation is printed from a solid block of a size covering as much of the already printed job as may be required to be glossed. It is very quick drying, and only as much must be put on the ink-table at one time as will be sufficient for a few minutes' use.

Transparent Paint for Glass is made by rubbing the colours chosen in a size made of Venice turpentine, two parts, and spirit of turpentine, one part.

Dryer for Printing Ink.—Mix 6 oz. of balsam of copaiba with one quart of spirits of turpentine; add a little to the printing ink when necessary. This dryer will make the ink work freely, and, in addition, will brighten the colour.

Lithographic Writing Ink.—Tushe is the name for this. Almost all the tushe in the market cracks easily and does not flow well from the pen upon zinc. This may be remedied by preparing a tushe, in the usual manner, of 8 parts white beeswax, 5 parts Castile soap, 15 parts shellac, 1 part mastic (in drops), 10 parts dragon's blood, 6 parts tallow, and 1 part lampblack. Form into sticks, and when using dissolve one part of this tushe in ten parts of *boiling* water. This tushe has proved the very best for work upon zinc.

Rubber Varnish.—The waste scraps of vulcanized rubber—a mixture of rubber and sulphur—which dealers in hard rubber goods can supply in abundance, will furnish an excellent and quick-drying varnish. Its colour can be varied from a golden yellow to the deepest brown. It sticks very well to metals, and can be employed on electric apparatus. The clippings are put into a deep earthenware pot, covered with a tight lid, and set upon hot coals. At the end of five minutes take the pot off from the fire and see if the material is melted. While the pot is on the fire take care not to lift the lid up, because the vapours which would be thrown off take fire easily. After the rubber is all melted, so that it can be poured out and there are no more whole pieces to be seen, pour it into a flat basin. The basin should be rubbed with grease beforehand, and after the mass is cooled it is readily detached. Then break into pieces, put it into a large bottle, pour on some benzole and rectified spirits of turpentine, and shake the mixture up several times. The solution being complete, pour out the liquor to get rid of the impurities and hardened rubber which remain at the bottom, and a very limpid, beautiful, and excellent varnish is obtained.

Glossy Writing Ink.—Any common writing ink can be made glossy by adding to it a little gum arabic or white sugar. If the latter is used, care must be had not to use too much sugar, else the mixture will be sticky when dry; and if too much of either gum or sugar is used, the ink will become too thick to flow well.

White Ruling Ink.—Chinese white, mixed with water containing enough gum arabic to prevent the immediate settling of the substance, makes a good white ink for ruling purposes. Magnesium carbonate may be used in the same way. Both must be reduced to impalpable powder.

Indelible Ink.—The appended formula is for a crimson marking ink:—Dissolve 1 oz. nitrate of silver and 1½ oz. carb. soda in crystals, separately, in distilled water. Mix the solutions, collect, and wash the precipitate on a filter; introduce the washed precipitate, still moist, into a Wedgwood mortar, and add to it tartaric acid 2 drs. 40 grs., rubbing together until effervescence ceases. Dissolve carmine, 6 grs., in liquor ammonia (0.883), 5 oz., and add to it the tartrate of silver, then mix in white sugar, 6 drs., and powdered gum arabic 10 drs., and add enough distilled water to make 6 oz. This ink is red when written with, but becomes black.

Preserving Printers' Inks.—A thin skin soon forms on the surface of inks only occasionally used, and, if not very carefully removed before using, small particles are easily distributed by the rollers in the ink, thus producing unclean work. In order to avoid this, every time ink is being taken from the can the surface should be made perfectly smooth again, and the particles adhering to the tin carefully removed, so as to diminish those parts liable to dry up. As soon as the required quantity of ink has been taken from the can, the latter should be covered with a sheet of parchment previously soaked in oil and well dried on both sides. Care should be taken that the ink is completely covered by the parchment. Then the can should be hermetically closed and wrapped in paper. Thus the ink can be preserved in good condition for a long time.

Photo-lithographic Inks.—The ingredients for photo-lithographic inks are :—Printing ink, 23 parts ; wax, 50 parts: tallow, 40 parts ; colophony, 35 parts ; oil of turpentine, 210 parts ; Berlin blue, 30 parts.

Liquid for Brightening Common Qualities of Coloured Inks.—Take white of fresh eggs, but apply very little at a time, as they dry hard, and are likely to take away the suction of the roller if used for any length of time.

Ink for Bronze and Metal.—A mixture of equal parts of Canada balsam and copal varnish, or some japanners' gold size, may be employed as an admixture for temporary gold ink. A drop or two applied to the slab or roller will give sufficient tenacity to the ordinary ink to enable proofs to be pulled. For regular employment, the ink should contain some strong or middle varnish and some dryers. The following is highly commended by some printers :—

Middle Varnish	2 parts.
White Wax	1 ,,
Venice Turpentine	1 ,,
Patent Dryers and Burnt Umber, of each a sufficient quantity.	

Plate Transfer Ink.—A good plate transfer ink may be made with the following ingredients :—Take 4 ounces each of tallow, wax, soap, shellac, pitch, and lithographic ink, and melt the materials in the order named. Burn the first three for fifteen minutes, put out the flame, and then add the shellac. When the shellac is dissolved, add the rest. Continue the heat for another fifteen minutes without setting on fire. Let a piece get cold, and, if not found hard enough, continue the heat until the necessary degree of hardness is arrived at, which may be known by its breaking with a sharp sound.

Cardboard Enamel.—Take one pound of parchment cuttings, one quarter-pound of isinglass, and one quarter-pound of gum arabic in four gallons of water; boil in an iron kettle until the solution is reduced to twelve quarts : it is then removed from the fire and strained. The solution is divided into three parts of four quarts each; to the first portion are added six pounds of white lead ground fine in water, to the second portion are added eight pounds of white lead, and to the third are added six pounds of white lead. The sheets of paper or cardboard are stretched out upon flat boards, and brushed over with a thin coat of the first mixture with an ordinary painter's brush; the paper is then hung up to dry for twenty-four hours. After this the paper is ready to receive a coat of the second mixture, and again hung up to dry for twenty-four hours; the paper is again treated in the same way with the third mixture, and dried for twenty-four hours. After this it receives a high gloss, which is obtained by laying the work face downward on a highly-polished steel plate, and then passing both with great pressure between a pair of powerful rollers.

Aniline Black for Printing on Cloth.—The following is recommended :—(*a*) 25 parts aniline oil are mixed cold with 20 parts nitric acid at 38° R. (*b*) 50 parts starch, 30 tragacanth, 35 acetic acid, and 20 potassium chlorate, are boiled with 200 parts of water. (*c*) A solution of ammonium vanadate equal to 2 per cent. of the aniline oil in (*a*) is mixed with (*a*) and (*b*). The mixture is printed, aged, and the goods worked off in a slightly alkaline bath of potassium chromate.

Engravers' Transfer Ink.—The compound used by wood engravers to make a transfer from a print on to a type-metal block consists of one ounce of caustic potash to half a pint of alcohol, made into a solution, with which the print is wetted for a few minutes. The type-metal block is then brushed over with Canada balsam, the picture put on face down, and the two run between rollers.

A Valuable Enamel for Artistic Purposes may be prepared from a mixture of thirty parts, by weight, of salt-petre, ninety parts of silicic acid, and 250 parts of litharge. Drawings can be made upon this enamel as upon paper, and the characters can be burnt in by means of a muffle in less than a minute. It can also be employed in the preparation of photographs without the use of collodion. For this purpose a mixture of ten parts of gum, one of honey, and three of bichromate of potash, well filtered, is dried upon the enamel and exposed in the camera, the image being then developed by dusting over it a powder of ten parts, by weight, oxide of cobalt, ninety of finely pulverized iron scales, a hundred of red lead, and thirty of sand ; the chromate is decomposed by immersion in a slightly acidulated bath. When washed and dried the enamel is melted by placing it upon a piece of clean sheet iron, coated with chalk, and the photograph glazed upon the enamel is then brought to view.

To prevent Spreading of Writing Ink on Tissue

use a fair proportion of gum arabic.

China (or India) Ink.—This is the ink universally used throughout the Chinese empire for writing with a brush, or painting, on one side only of the paper in general use in China. It is rubbed down on a porcelain slab, or palette, and used with a fine camel-hair brush for painting upon the soft, smooth, flexible rice paper employed by Chinese artists, and which is such a peculiarly characteristic product of those ingenious orientals. China ink, so far as is known, was the earliest artificial ink made, and it is still in general use in China and Japan. It has also been used in India so long as to have obtained the name of Indian ink. In Europe it is employed to some extent by architects and engineers, and also by artists, who use it for designs or sketches in black and white, for which it possesses the advantage of affording various depths in shading according to its greater or less dilution with water. It is made of lampblack and size or animal glue, with the addition of perfumes or other substances not essential to its quality as an ink. It is prepared in the form generally of small sticks, which are formed in wooden moulds.

Ruling Inks.—The following is a good recipe and drier :— Sulphuric ether, one ounce ; refined spirits turpentine, one-quarter ounce : mix and use in heavy lines, but use sparingly.

Orange Ruling Ink.—To make this, mix red and yellow in proper proportions to the shade desired ; there is no set rule for quantity ; add a few drops of ammonia to set the colour.

Sea-green Ruling Ink.—To make this add blue indigo paste to picric acid yellow, sufficient to give the shade wanted ; add gall when using the same as other inks.

Ink Tablets.—The demand for ink tablets or powders is limited, though the form is extremely handy for carrying on a journey. Two receipts for preparing ink tablets are here given : 1. Extract of logwood, 500 parts ; alum, 10 parts ; gum arabic, 10 parts ; neutral chromate of potassium, 1 part. Dissolve the salts in 500 parts of water, add the extract of logwood and gum arabic and concentrate the mixture to the consistence of an extract. Then pour the mass out either into moulds or into a flat-bottomed dish, and cut it in pieces of suitable size, which may be enclosed in boxes or other receptacles. 2. Extract of logwood, 100 parts ; gum arabic, 10 parts ; indigo carmine, 5 parts ; neutral chromate of potassium, 1 part ; glycerine, 10 parts ; water, q. s. Proceed as in the preceding formula.

Indelible Ink for Rubber Stamps, which is said not to injure the rubber, is made of boiled linseed oil varnish 16 parts, best lampblack 6 parts, perchloride of iron 2 to 5 parts. It should not be used for metal type.

Coloured Writing Inks.—Those which are not fugitive—that is, which will not fade and into which it is safer that aniline colours should not enter—can thus be prepared :—

Red.—Four oz. ground Brazil wood, and 3 pints vinegar, boiled till reduced to $1\frac{1}{2}$ pint, and 3 oz. powdered rock alum added.

Purple.—To a decoction of 12 parts Campeachy wood in 120 parts water, add 1 part subacetate of copper, 14 parts alum, and 4 parts gum arabic ; let stand 4 or 5 days.

Violet.—Boil 8 oz. logwood in 3 pints water, till reduced to $1\frac{1}{2}$ pint ; strain and add $1\frac{1}{2}$ oz. gum, and $2\frac{1}{2}$ oz. alum.

Blue.—Two oz. Chinese blue, 1 quart boiling water, 1 oz. oxalic acid ; dissolve the blue in the water, and add the acid ; it is ready for use at once.

Green.—Two oz. verdigris, 1 oz. cream of tartar, $\frac{1}{2}$ pint water ; boil till reduced one-half, and filter.

Black Writing Ink.—Take six ounces of the best gall nuts, and pound them in a mortar or otherwise; take four ounces of logwood, and let it be cut or ground into very small pieces; these, mixed with four quarts of rain or river water, must be boiled together until half diminished. Then take two ounces of copperas made into a powder, and three ounces of gum arabic; let these be also mixed and strained through a linen cloth. After this mixture has stood a few hours it may be written with. The ink thus prepared is very fine, and makes the writing appear beautiful and shining. Another old-time recipe is the following:—Take three quarts of rain water and sixteen ounces of pounded gall nuts; boil these on a slow fire until the liquor has diminished two-thirds, then throw it into two ounces of gum arabic, which has been already dissolved in half a pint of vinegar, then add six ounces of powdered copperas; let this mixture boil for two hours longer, and then it may be bottled for use.

Ink Slabs for Colour Work.—It has been suggested that all ink-tables on machines used for colour work should be nickel-plated. Iron "kills" bright colours, and renders tints dull and lifeless. The cost of the nickeling is trifling, and once done lasts a lifetime. Porcelain, litho, stone, marble, slate, and glass are all good surfaces to use for colour slabs.

Substitute for Gold Size.—Few printers there are who have not had trouble with the gold size regularly furnished by ink-makers. Not only has it an invariable tendency to dry on the rollers and on the forme, but it too frequently dries on the sheets before the bronze can be dusted on. The result is an effect that suggests the employment of the very poorest grade of bronze, even where the finest of gold is used. Again, the regular stock size is more than likely to fill up the finer lines of engravings and shaded letters wherever employed, giving a dirty, muddy effect, as though too much ink had been used. A moment's reflection will show how inevitable this result must be where such an extra glutinous and quick-drying pigment is constantly thickening and hardening on the delicate hair-lines of a job, necessitating frequent stops and washings. This is a serious drawback to fine work, but it can be easily remedied by the employment of a substitute which does away with the above objections. It is made as follows: Take three parts of lemon yellow ink and one part of No. 2 varnish, and mix them well; add about one-twentieth of the above quantity of copal flock varnish, and mix perfectly. The peculiar quality of this size is that it can be run all day without washing it—an impossibility with the other; it never thickens the hair-lines of a job, works freely on and off the rollers, and allows sheets to lie longer without bronzing than any other method. The bronze shows less liability to smut or spread, and altogether cleaner and brighter work is produced, showing fully covered lines, instead of parts that seem only half bronzed. This size being slow to set, gives the bronzer a better chance to cover his job, but for this very reason holds more firmly and permanently every grain that is spread upon it.

As an Ink Extractor, freshly made, use saturated solution of chloride of lime, one ounce; acetic acid, one ounce; and as an extractor for black ink use a weak solution of oxalic acid.

Can for Printers' Ink.—The following is a description of an improved one of American origin. The head of the can is made integral with the body, an egress for the ink being provided by a nipple, which is situated in the centre of the can head, and may be closed by a screw-cap. A movable bottom—described in the specification as a "convavo-convex follower"—is adapted to slide longitudinally within the body of the can, and is surrounded by annular packing which prevents leakage without restricting motion. If the screw-cap be removed and the bottom of the can pressed in, the ink will be forced from the nipple in the same manner that paints are exuded from the collapsible tubes which contain them. This can possesses the two principal virtues of collapsible tubes: the manner of ejecting its contents, and the protection it affords its contents against dust. It differs from collapsible tubes in that, owing to its rigidity, it preserves its form, and may be refilled when empty.

Preserving Ink, Gum, or Paste.—It is said that a quarter per cent. of formic acid added to ink, gum, paste, and such other articles, will keep them as fresh as possible.

A Hint for Gold Printing.—To prevent gold-leaf or bronze adhering to the surface beyond the outline of the sizing, pounce the whole of the surface after sizing with whiting, or lay on with a soft brush whiting mixed with water, brushing off the superfluous powder when the water has evaporated. The varnish or gold size may be distinctly seen over this whitish ground as the gilding progresses.

Letterpress and Lithographic Inks.—There is a great difference between lithographers'and letterpress printers' inks, whether black or in colours. Printers can use lithographers' inks, and would do well to make use of them, but lithographers should never attempt to use printers' inks. Lithographers' inks, manufactured by reliable houses, are certainly the best for such purposes, for the reason that these inks have to be ground and mixed with suitable varnishes in order to obtain a solid and sharp printing, and a knowledge of this is necessary to manufacture a suitable ink, while, in the case of printers' inks, manufacturers are not so careful, and are satisfied if the printer is able to obtain a smooth impression. To do the exact thing it is necessary to know just what materials we are working with, and particularly in the case of inks we must know the ingredients, so that we can tell what chemical actions will result, and also how they will work upon the stone in the lithographic process, and with the paper used for printing upon.

Facilitating Bronze Printing.—Calcined magnesia rubbed on a job will allow of bronze being printed over a colour without adhering to it, but the colour should be as dry as possible before applying the magnesia.

The Sticking of Poster Sheets.—Poster ink, coloured or black, has a tendency to make sheets stick together as though they had been glued. Mix a small quantity of soft soap amongst it—this will generally put matters right.

Thinning Ink.—A printer of large practical experience says that he thins his ink with spirits of turpentine, and works it with dammar varnish previously thinned with raw (not boiled) linseed oil. The use of turpentine sets off, in the drying properties of the ink, the use of raw linseed. He has had, he says, most trouble with the red and green inks, both of which have been treated successfully in the manner described.

Stiff Ink.—If ink should pull the face of paper, owing to frost or to the paper itself having a soft, spongy, or enamelled face, put a drop or two of oak varnish on the slab, and run up. If very frosty weather, it may be found necessary to heat the slab. Some firms keep a small gas jet arranged under each slab for these emergencies.

Hint in using Copying Printing Ink.—Printers who use this should have a small bottle of glycerine on hand. If the ink does not take apply the glycerine on the ink-plate with the tip of the finger until the trouble is removed. It is absolutely necessary in using copying ink to have rollers and ink-disk perfectly clean and free from all other inks.

Reflection of Coloured Inks.—Mix a little of the white of a fresh egg with coloured inks at the time of use. This increases their reflection, and, at the same time, gives a siccative.

The Care of Printing Inks.—Messrs. Fleming and Co., printing ink makers, say:—The great difficulty that printers have to contend with is getting an ink which will print a dense enough colour, and yet will dry sufficiently quick, so as not to soil the hand shortly after it leaves the press. Inks which have coal-tar colours for their basis are, as a rule, affected by water and varnish (being soluble and the colours running); whilst light and exposure to weather have a most destructive effect. Vermilion inks, when printed from lead or iron types, cause sulphur combinations, which change the colour of the ink. It is needful, when used in machines with brass cylinders, to cover same with a coating of copal or dammar varnish. Caution is needful in using tint inks upon tinted paper to

K

prevent inharmonious combinations. By covering the top of ink in can, when opened, with oiled paper, or paper dipped in glycerine, skinning is prevented. For ultramarine inks, use oiled paper only, as glycerine affects this colour. In any special contracts it is advisable to send to the ink manufacturer a sample of paper used, and, where practicable, a small sample of the ink in ordinary use. As the quality of paper has much to do with success or non-success of an ink, the following observations will be of interest:—1st. Notice whether white or brownish; a brown tint in paper makes ink look brown. 2nd. If porous or glazed. If you press your tongue against it when porous, it will immediately get soft from absorbing the moisture, and become semi-transparent. Porous paper is, of course, less transparent, and with a strong ink is more apt to be torn; it requires, therefore, thinner ink; is generally printed upon in a damp state, less ink being required.

A Simple and Effective Lye.—Table salt, 2 oz.; unslacked lime, 2 lbs.; washing soda, bruised, 2 lbs.; put together in 3 gallons of soft water, stirred well. When settled ready for use, pour the liquor off, and throw the sediment of the lime away.

A Peculiar Ink Plant.—There is in New Granada a plant, *Coryaria thymifolia*, which might be dangerous to our ink manufacturers if it could be acclimatized in Europe. It is known under the name of the ink-plant. Its juice, called *chanchi*, can be used in writing without any previous preparation. The letters traced with it are of a reddish colour at first, but turn a deep black in a few hours. This juice also spoils steel pens less than common ink. The qualities of the plant seem to have been discovered under the Spanish administration. Some writings intended for the mother country were wet through with sea-water on the voyage; while the papers written with common ink were almost illegible, those with the juice of the ink-plant were quite unscathed. Orders were given in consequence that this vegetable ink was to be used for all public documents.

Type-writing Inks.—Some of these are fugitive: that is, they will not stand exposure to the strong sunlight. Purple, red, and green, if exposed for a great length of time, will fade. Red will fade out completely, and for that reason very few red ribbons are made. Purple being a colour very much liked it became a question with manufacturers how to render it permanent, and this has been done by simply adding carbon, which is indestructible, to the purple ink, making what is known as " black-copying purple." This ribbon prints black, but, of course, in making a letterpress copy the copy is purple. The original takes a purplish tinge. The copy, if exposed, is liable to fade, but it is as permanent as the greater number of the ordinary fluids. The original is permanent, being rendered so by the carbon. There is an indelible ribbon, writing a deep blue-black and making a strong blue copy. This is perhaps the best for all purposes, as both the copy and the original are absolutely permanent. Ordinary printing ink may be used if a black is required.

Another Ink for Type-writer Ribbons.—To make this, take vaseline of high boiling point, melt it in a water bath or slow fire, and incorporate by constant stirring as much lampblack as it will take up without becoming granular. Remove the mixture from the fire, and while it is cooling mix equal parts of petroleum, benzine, and rectified oil of turpentine, in which dissolve the fatty ink introduced in small portions by constant agitation.

New Printing Varnish.—The composition is: two parts of painters' terebene ; one part of linseed oil, and one part of Canada balsam turpentine. The painters' terebene may be manufactured by boiling half a pint of linseed oil and two drachms oxide of lead together for an hour, with incessant stirring, afterwards adding a few drops of acetic acid. To impart the glaze, instead of the Canada balsam, there may be used two pounds of pale, hard copal, one pint of linseed oil, and three pints of turpentine, dissolved together. By the use of the above, the inventors claim the inks to be rendered impervious to sun, rain, and bill-posters' paste for a considerable time, as well as increased brilliancy.

Powdered Red in Lithography.—The ink is composed of strong varnish, medium varnish, and Baltimore yellow, well ground. The following proportions may be taken as the basis:—Strong and medium varnish, and an equal quantity of Baltimore yellow. Well mix the varnishes, add the yellow, and grind the whole thoroughly. Pressure being sufficiently heavy, the yellow should appear very distinct. Its shade is by no means harmful to the red. On the contrary, the yellow strengthens the red. When the red powder tints the white parts of the paper, a small liqueur glass of linseed oil should be used to a plate of powder. Mix with a piece of card until the colour has entirely absorbed the fatty body. The addition of this quantity of oil, rendering the powder fatty, prevents its soiling the paper.

Violet Copying Ink.—The following formula is for the preparation of a violet copying ink:—40 grams extract of logwood, 5 grams oxalic acid, and 30 grams alum are dissolved in 800 grams rain water and 10 grams glycerine, and the whole allowed to stand for twenty-four hours. The ink is then brought to the boiling point in a copper vessel and 50 grams wood vinegar are added, and, after standing for a while, the ink is filled into bottles.

Hints about Printers' Inks.—For all commercial work, printed on writing-paper, use ink with good body—a short, thick ink covers better on writing-paper. For circulars or other work printed on super-sized and calendered book the same ink will of course answer, but a cheaper quality will do as well. For jobs on print paper the thinnest ink is best. Short ink and hard rollers, thin ink and soft rollers, go best together. Where there are solid surfaces to cover, thick ink is best, because it covers better. For the same reason short ink is cheaper than the tacky article. Never try to print cuts or fine work on good paper with a poor quality of ink. It does not pay. Use the thick-bodied, short ink for such work, especially on a platen press. Wash up at least once a day, even on long runs. The ink will become clogged with dust, and good work is then impossible. Always use the best of

coloured inks, except for poster work. They are cheaper in
the end. In opening a can of coloured ink of which little is
required, do not pull the skin off the top ; break it at the side,
take out what is wanted, and immediately replace the skin.
If you take the skin all off it will form again, and the ink will be
wasted. Keep all ink-cans well covered. Dust will ruin
any ink.

Imitation India Ink.—A colour apparently identical
with India ink can be produced by the action of sulphuric
acid on camphor. An excess of camphor should remain some
twenty-four hours in strong sulphuric acid ; it then results in
a gelatinous mass of a slightly reddish colour. This, when
heated, effervesces, gives off fumes of sulphurous acid and
turns intensely black. By evaporation the superfluous sulphuric acid and camphor (for there remains an excess of both,
the weakened acid not acting on the camphor) can be driven
off. The remainder, when applied to paper as a paint, appears
to be India ink. When dissolved in water it remains an
indefinite time without precipitating. It appears to be dissolved, not held in suspension.

To make Gold Bronze.—Melt two parts of pure tin in
a crucible and add to it, under constant stirring, one part of
metallic mercury, previously heated in an iron spoon, until it
begins to emit fumes. When cold, the alloy is rubbed to
powder, mixed with part each of chloride of ammonium and
sublimed sulphur, and the whole enclosed in a flask or retort
which is embedded in a sand bath. Heat is now applied until
the sand has become red-hot, and this is maintained until it
is certain that vapours are no longer evolved. The vessel is
then removed from the hot sand and allowed to cool. The
lower part of the vessel contains the gold bronze as a shining
gold-coloured mass. In the upper part of the flask or retort,
chloride of ammonium and cinnabar will be found.

Recipe for Silver Ink.—This is white gum arabic, one
part ; distilled water, four parts ; silicate of soda in solution,
one part. Triturate with the best silver bronze powder enough
to give the required brilliancy.

Red Marking Ink.—A German formula which is said to be unaffected by either soap, alkalies, or acids, is the following: Enough finely pulverized cinnabar to form a moderately thick liquid is very intimately mixed with egg albumen, previously diluted with an equal bulk of water, beaten to a froth and filtered through fine linen. Marks formed on woven tissues with this liquid, by means of a quill, are fixed after they have become dry, by pressing the cloth on the reverse side with a hot iron. The ink will keep in well-closed bottles for a long time, without separation of the suspended cinnabar.

Recipe for Blue-Black Writing Ink.—The following dependable recipe is published in response to a request made by several inquirers. To 1 lb. of bruised galls add a gallon of boiling water; 5½ oz. copperas in solution; 3 oz. gum arabic in solution; and a few drops of carbolic acid. Finally add a strong solution of fine Prussian blue, in sufficient quantity to turn the ink a blue-black when written with. This ink afterwards turns a jet black, and cannot be erased either by acids or alkalies without the destruction of the paper.

Another Black Marking Ink.—A good serviceable marking ink for use on linen without preparation is made as follows: Dissolve separately one ounce of nitrate of silver and one and a half ounce of best washing soda in distilled or rain water. Mix the solutions and collect and wash the precipitate in a filter; whilst still moist rub it up in a marble or Wedgwood mortar with three drachms of tartaric acid; add two ounces of distilled water, mix six drachms of white sugar, ten drachms of powdered gum arabic, half an ounce of archil, and water to make up six ounces in measure.

White Varnish for Paper.—The following is the same as that used for foreign wood toys, and is composed of tender copal, one and a half ounce; camphor, 1 ounce; alcohol of 95 per cent., 1 quart, to which, when dissolved, is added mastic, 2 ounces; Venice turpentine, 1 ounce. The whole is then dissolved and strained. This varnish is extremely hard.

Clear Shellac Varnish.—To get an absolutely clear solution of shellac has long been a desideratum, not only with microscopists, but with all others who have occasional need of the medium for cements, etc. It may be prepared by first making an alcoholic solution of shellac in the usual way; a little benzole is then added and the mixture well shaken. In the course of from twenty-four to forty-eight hours the fluid will have separated into two distinct layers, an upper alcoholic stratum, perfectly clear, and of a dark red colour, while under it is a turbid mixture containing the impurities. The clear solution may be drawn off.

The Varnishing of Paper.—To make size for wall-paper, break some glue up small, put it into a pail and cover the glue with water, and allow it to soak for ten or twelve hours; then add more water and boil until dissolved. Strain it through a muslin cloth, and try the size on a piece of paper. If it glistens, it is too thick; then add water. If it soaks into the paper it is too thin. Be careful, especially in the first coat, to bear very lightly upon the brush, and have plenty of size to flow freely from it, otherwise you may damage the paper. Give two coats of this, and when dry varnish with pale varnish, which should be applied very briskly, and leave off at the flow.

WAREHOUSE WORK AND STATIONERY.

TO Make Paper Tough.—A plan for rendering paper as tough as wood or leather has been recently introduced. It consists in mixing chloride of zinc with the pulp in the course of manufacture. It has been found that the greater the degree of concentration of the zinc solution, the greater will be the toughness of the paper. It can be used for making boxes, combs, for roofing, and even for making boots.

On Judging Paper.—When writings are yellow wove, azure laid, or blue, they are frequently darker on one side than on the other. *When the paper is darker in colour on the right side it is hand-made; when darker on the wrong side it is machine-made.* This method of distinguishing cannot be taken as conclusive without corroboration from other sources, for means are now taken on the machines in some mills to counteract the subsidence of the blue pigment. It may happen that this rule fails, owing to a high finish having been imparted to the right side of a machine paper, by which it would become darker on that side. An unsought watermark is impressed by the wire cloth on all papers made by machine. It is indelible. No after-process of surfacing can obliterate it. *It is always present in laid machine-papers, but is never found in laid hand-mades.* The mark, which is a facsimile of the wire cloth, owes its existence principally to the suction of the exhaust boxes; and according to the intensity of their action on the fibre, so does the texture-like impression in the finished sheet vary. Hence it is more distinct in some papers than others. With a very fine wire,

light action in the exhaust boxes, and a good finish afterwards, these wire marks get so faint that they are only discernible on looking through the sheet at a strong light. Two
marks, laid and wove, in the same sheet, are an infallible indi-
cation of the sample being made by machine.

Enamelling Cardboard and Pasteboard.—Dis-
solve ten parts of shellac in a sufficient quantity of alcohol, and
add ten parts of linseed oil. To each quart of the mixture add
also about one-fourth of an ounce of chloride of zinc. The
board may be immersed in it or the solution applied with a
brush. The board is thoroughly dried and the surface is
polished with sand-paper or pumice before applying this preparation.

Paper Pulp from Cotton Stalks.—A writer in the
" Scientific American " says that several samples made from
the hulls and stalks of the cotton plant have lately been on
view at Atlanta, Georgia. The pulp is as white as snow, and
it is said that it can be converted into the finest writing paper.
The ligneous substances of the hulls and stalks are removed
by a new process. Fifty per cent. of the fibres are extracted
from the hull, which has hitherto been used either for fuel in
the mills or for fertilizing purposes, and 38 per cent. is obtained
from the stalks, which have generally been allowed to rot in
the fields. If the process proves successful, the value of these
comparatively useless products will be increased tenfold.

Gummed Paper. — The tendency of paper when
gummed (in the case of postage stamps, labels, etc.), to curl
up is very tiresome, and much waste is often caused through
tearing. It is said that this evil may be avoided by adding a
little salt, sugar, and glycerine to the gum—very little of the
latter, however, because otherwise the gum does not dry
thoroughly. The gummed paper, also, must not be dried in
too great a heat. Another peculiarity of gummed paper is its
greater liability to curl up the thicker it is. The thinnest
paper possible under various circumstances should therefore
be used.

Waterproof Luminous Paper, which will shine in the dark, is made as follows :—40 parts paper stock (pulp), 10 parts phosphorescent powder, 10 parts water, 1 part gelatine, and 1 part bichromate of potash.

To Take Creases out of drawing paper or engravings, lay the paper or engraving face downward, on a sheet of smooth, unsized, white paper ; cover it with another sheet of the same, very slightly damped, and iron with a moderately warm flat-iron.

Bronzing Paper.—Dissolve gumlac in four parts by volume of pure alcohol, and then add bronze, or any other metal powder, in the proportion of one part to three parts of the solution. The surface to be covered must be very smooth and carefully polished. The mixture is painted on, and when a sufficient number of coats have been given, the object is well rubbed. Another method is to coat the object with copal or other varnish, and when this has dried so far as to become " tacky," dust bronze powder over it. After a few hours the bronzed surface should be burnished with a burnisher of steel or agate.

To Remove Grease Spots from Paper.—The following is a recipe for removing grease spots from paper :— Scrape finely some pipe clay on the sheet of paper which is to be cleaned. Let it completely cover it, then lay a thin piece of paper over it, and pass a heated iron on it for a few seconds. Then take a perfectly clean piece of india-rubber and rub off the pipe clay. In most cases one application will be found sufficient, but if it is not, repeat it.

Insensible Paper.—A new process of rendering paper insensible to the action of water and atmospheric changes has been patented in France. A sheet of paper is covered on the wrong side with a thin layer of gutta percha, which is afterwards spread with paper, linen, thin pasteboard, or similar matter. The whole is heated and pressed. Under the influence of heat the gutta percha becomes softened and unites firmly the two surfaces between which it has been placed. This prepared paper is of great value in art work.

Papier-maché Goods.—Papier-maché is pulp paper
moulded into form, and if required waterproof, sulphate of iron, quicklime, and glue (or white of egg) are added to the pulp, and if incombustible, borax and phosphate of soda are added. Tea trays and other light goods are made by pasting or gluing sheets of paper together, and then submitting to powerful pressure. The articles are afterwards japanned, and are then perfectly waterproof. The refuse of cotton and flax mills may be used instead of paper pulp.

Transparent Paper.—The following methods for ren-
dering paper transparent are given. Using castor oil answers as well as any other method, the best recipe being :—Of castor oil five parts and of ether one part ; place the paper upon a sheet of glass, and spread the solution thickly over it ; well warm it till the oil has thoroughly soaked into the paper ; when cool remove the superfluous oil and again warm. Another method adopted is by using Thomas' india-rubber solution, two parts, dissolved with two parts Canada balsam in three parts pure benzole, and rubbing well in with a piece of cotton wool till thoroughly soaked and dry. Passing through melted paraffin wax is also an excellent method. This must be effected at such a temperature as to enable it to thoroughly penetrate the paper. Better *not* to iron, as so often recommended, but simply to warm, and with a piece of soft cloth take off the superfluous wax. A process by no means easy, but which we have ourselves carried out with great success, is the following :—Gum dammar twenty parts, and gum elemi five parts, dissolved in a hundred parts of benzole. Pour into a flat dish, place the paper in one sheet after another, and allow it to remain for about five minutes ; then remove and hang up to dry. Benzole must be constantly added to the solution in consequence of its speedy evaporation.

Cloth Finished Paper.—A cloth finish is given to paper by applying to, or laying upon, opposite surfaces of paper, pieces of cloth, subsequently subjecting the cloth and paper to pressure between smooth rollers or other smooth surfaces, and in finally removing the cloth from the paper. The paper or the cloth may, if desirable, be moistened to facilitate the impress of the surface of the cloth into the paper; but this will not be necessary. The impress of the surface of the cloth into the paper may be done, in the manner described, either before or after calendering the paper, or even during the process of calendering. A name or designating mark or ornament may be produced in the paper by delineating it upon the cloth by stitching, or in any other manner which will give it the necessary projection.

Baskets of Paper.—Almost any shaped baskets and workcases can be made of plaited paper. White, brown, or newspaper may be utilized for this purpose. To make a white basket, cut into narrow strips, three-quarters of an inch wide, thin cartridge paper; double each strip lengthwise into two and plait it. When you have plaited a sufficient quantity stitch the pieces together into the shape you wish the basket to be, using a small basin, jar, or even another basket to assist in shaping the work. When the basket is stitched into shape, and the handles put on, give it two coats of hot gelatine and water, and when thoroughly dry, varnish. Baskets of newspaper and brown paper must be painted with oil colour and afterwards varnished.

Blacking Skins.—The "skins" used to wrap up paste blacking consist simply of manilla tissue, or strong small hand similar to that used for paper bags, well saturated in oil (after printing) and hung up to dry.

Ornamentation on Paper.—A new method of producing designs and patterns in colours upon the surface of paper consists in the use of one or more rollers or cylinders of elastic material, such as vulcanized india-rubber, filled with compressed air. The diameter of these rollers or cylinders is determined by the pressure of the air. These rollers or cylinders are closed at the ends, and supported by an axle in a frame. Any suitable design or pattern is produced upon the surface of these rollers or cylinders. This may either be done by cutting the design or pattern out of the surface, or by cutting the ground out of it; or it may be produced by a mould. The rollers revolve by contact with the advancing paper, and projecting portions of its engraved surface take up more or less of the colour and the designs or patterns in the moist colour on the paper. It is said that these cylinders will produce designs or patterns with very soft shading, which have not been produced hitherto by any mechanical contrivance.

Safety Envelopes.—In making these envelopes, that part of the envelope covered by the flap is treated with a solution of chromic acid, ammonia, sulphuric acid, sulphate of copper, on fine white paper. The flap itself is coated with a solution of isinglass in acetic acid, and when this is moistened and pressed down on the under part of the envelope a solid cement is formed, entirely insoluble in acids, alkalies, hot or cold water, steam, etc.

Preparing Paper for Copying Purposes.—A new method or process for treating paper so as to render it permanently moist for copying purposes has been devised. In preparing the paper one pound of the salt known as the "chloride of magnesium" is dissolved in a moderate quantity of cold or warm water, and it is ready for use. From half a pound to a pound of water to the pound of chloride of magnesium is said to be the most desirable amount, but more or less can be used according to circumstances. The chloride of magnesium may be dissolved in other liquids, but water is said to be equally good. Apply this solution to the sheets of ordinary copying-paper, whether in book form or otherwise, in any usual and well-known manner, and preferably by applying the compound to cloth pads well saturated with the liquids, and then place the pads between any suitable number of leaves; then apply a pressure, at first very moderate, until the absorption of the paper is complete ; then remove the cloth paper and apply under the press a strong pressure, and the books or sheets of paper so treated are ready for copying purposes, the use of the solution of chloride of magnesium being the radical or base of this invention. In all cases use copying inks or fluids, which are preferable. Paper prepared by this method will remain permanently moist at any ordinary temperature, and if made dry by any extraordinary heat will regain its moisture upon being subjected to the common temperature.

Blasting Paper, as made by an Austrian firm, consists of unsized paper coated with a hot mixture of 17 parts of yellow prussiate or potash, 17 of charcoal, 35 refined saltpetre, 70 of potassium chlorate, 10 of wheat starch, and 1,500 of water. After drying it is cut into strips and rolled into cartridges.

Set-off Paper.—A capital one may be made by lightly rubbing with glycerine. This is preferable both to oil and paraffin wax.

Artificial Parchment.—A strong artificial parchment, impermeable by water, and capable of serving for the diaphragm in osmotic operations on solutions of impure sugar, etc., is made as follows :—The woollen or cotton tissues are freed by washing from the foreign substances, such as gum, starch, etc., which may cover them. They are then placed in a bath slightly charged with paper pulp; and to make this pulp penetrate more deeply, they are passed between two rollers, which slightly compress them. The principal operation consists in steeping the product for a few seconds in a bath of concentrated sulphuric acid, after which it undergoes a series of washings in water and ammoniacal liquor, until it has lost all trace of acid or base. It is then compressed between two steel rollers, dried between two others, covered with felt, and finally calendered, when it is fit for use.

Mill-Boards, or properly "milled" boards, are strong flexible boards made chiefly of old rope. They are so called because they are squeezed or rolled in the process of manufacture. The best sorts are made from the same material as brown paper—old tarred rope; besides this, old coal sacks are used, with admixtures of various fibres. Mill-boards are made in the same way as hand-made paper, that is, in a mould, to insure firmness and solidity. To give them the necessary smoothness, they are finished by being rolled or milled by powerful iron and steel rolls.

Paper Bottles, capable of fully withstanding the effect of alcoholic fluids, and unbreakable, are manufactured from 10 parts rags, 50 parts wood, and 40 parts straw stocks. Both sides of the sheets are covered with a mixture of defibrinated blood and powdered lime. Ten thicknesses of this material are placed one on another, and are then pressed in hot metal moulds until they assume the form of half a bottle, The two halves are then united under the influence of heat and pressure into a perfect bottle, *i.e.,* they are to a certain extent welded together.

Hanging Paper for Damp Walls.—This method consists of coating a lining paper on one side with a solution of shellac spirit, of somewhat greater consistency than the ordinary "French polish," and then hanging it with the side thus treated to the damp wall. The paper-hanging is then performed in the usual manner with paste. Any other resin that is equally soluble in spirits may be used in place of the shellac. According to representations, this process is found equally effective in preventing the penetration of dampness.

To Soften Hard Paper.—It is a well-known fact that hard paper will become smooth and take the ink readily when a little glycerine is added to the water used for wetting purposes. But it may be less known that the ink will also dry very quickly on paper wetted with glycerine water. Posters with large and full-faced types will be dry in a quarter of an hour, whilst the drying process, when the printing has been done on paper simply water wetted, will require hours.

" **Retree.**"—The origin of the word "retree" cannot be traced farther back than the reign of the Great Napoleon, prior to the Peninsular war. He desired that the soldiers in every regiment should be of uniform height, and also of uniform physical formation, so that each man should appear equal to his fellow in the proportions of manhood. The words *trier de soldats* were thus first used, which meant "Pick of the soldiers." A subordinate officer first arranged those in line who, according to his judgment, were of equal height and physical proportions. Next, the colonel of the regiment inspected and further picked out those unsuitable for the crack regiments. Then Napoleon himself would pass down the lines, and further weed out those he considered not quite up to the mark. The weeded men were called *retrié*, but the French pronunciation allows of the Anglicized *retree*, as we now spell the word; the absolute meaning being "thrice picked." The English word *retree* is now well known and acknowledged in the trade. Fine papers are turned over sheet by sheet, and the finest qualities are divided into three parcels, only clean perfect sheets being passed for good paper. Those sheets which contain spots, or are otherwise imperfectly made, are separated from the well-authenticated sheets, and the broken make is called *retree*.

Papier-maché Covering for Floors.—A new and desirable process is described as follows:—The floor is thoroughly cleaned. The holes and cracks are then filled with paper putty, made by soaking newspapers in a paste made of wheat flour, water, and ground alum, as follows :—To one pound of flour add three quarts of water and a tablespoonful of ground alum, and mix thoroughly. The floor is then coated with this paste, and then a thickness of manilla or hardware paper is put on. If two layers are desired, a second covering of manilla is put on. This is allowed to dry thoroughly. The manilla paper is then covered with paste, and a layer of wall paper of any style or design desired is put on. After allowing this to thoroughly dry, it is covered with two or more coats of sizing, made by dissolving one-half pound of white glue in two quarts of hot water. After this is allowed to dry the surface is given one coat of "hard oil finish varnish." The process is represented to be durable and cheap, and, besides taking the place of matting, carpet, oil-cloth, or like covering, makes the floor air-tight, and can be washed or scrubbed.

Antiseptic Paper is prepared by melting five parts of stearine, in which two parts of carbolic acid are well stirred. Five parts of paraffin are then added to the mixture. The whole is stirred as it cools, and put on the paper with a brush.

To Make Postal Tubes.—The tubes now so largely used are made by rolling a sheet of the paper selected—cartridge or stout rope-brown—on a cylinder of wood of the required length and diameter of inside required. If you want a tube two feet long cut the paper to that width, and about a yard in length ; paste the sheet evenly all over, and then proceed to roll tightly on the wooden cylinder. In about an hour the tube will be dry, and the cylinder may be pushed out.

Paper Windows.—One of the most remarkable uses to which paper has of late years been put is the manufacture of xylonite—a substance which at the will of the manufacturer may be made in imitation of horn, rubber, tortoiseshell, amber, and even glass. The uses are almost infinite to which xylonite is adaptable, but the most extraordinary, perhaps, is the manufacture of cathedral windows ! The discovery was made several years ago by an Englishman named Spills, but it was not till some years after that a company was formed in London for its manufacture. The basis of xylonite is plain white tissue paper, made from cotton or cotton and linen rags. Being first treated with a bath of sulphuric and other acids, the paper undergoes a chemical change. The acid is then carefully washed out, and the paper treated with a preparation of alcohol and camphor. After this it assumes an appearance very much like parchment. It is then capable of being worked up into plates of any thickness, rendered almost perfectly transparent, or given any of the brilliant colours that silk will take.

L

Sizes of Envelopes.

	In Half.	In Three.
Queen	$3\frac{3}{4} \times 3$	... ...
Albert	$4\frac{1}{8} \times 3\frac{1}{4}$	... ...
Post	$4\frac{3}{4} \times 3\frac{3}{4}$	$4\frac{3}{4} \times 2\frac{3}{4}$
Large Post	$5\frac{1}{4} \times 4\frac{1}{4}$	$5\frac{1}{4} \times 3$
Demy	$5 \times 3\frac{7}{8}$	$5 \times 2\frac{3}{4}$
Medium	$5\frac{3}{4} \times 4\frac{1}{2}$	$5\frac{3}{4} \times 3$

Number of Cards contained in a Royal Board.

Thirds	96	Double Small ...	25
Broad Thirds ...	80	Double Large ...	16
Small	50	Quad Small......	12
Large	32	Quad Large......	8

Sizes of Tea and Tobacco Papers.

Tea :   1 *lb.* Foolscap.

    „ $\frac{1}{2}$ *lb.* Crown folio, or 14 × 11 in.

    „ $\frac{1}{4}$ *lb.* Demy 4to.

    „   2 *oz.* Demy 6to, or Crown 4to.

    „   1 *oz.* Demy 9mo.

Tobacco : $\frac{1}{4}$ *lb.* Crown 4to.

    „   2 *oz.* Crown 6to.

    „   1 *oz.* Crown 9mo.

    „ $\frac{1}{2}$ *oz.* Crown 12mo.

Number of Pages in a given Number of Quires.

1 Quire equal to 96 pages.		7 Quires equal to 672 pages.	
2 „ „ 192 „		8 „ „ 768 „	
3 „ „ 288 „		9 „ „ 864 „	
4 „ „ 384 „		10 „ „ 960 „	
5 „ „ 480 „		11 „ „ 1056 „	
6 „ „ 576 „		12 „ „ 1152 „	

Parchment.—Five dozen skins = 60 is called a Roll.

Equivalent Sizes of Writing and Printing Papers and Cartridges, shown at a glance.

Description.	Writings.	Printings.	Cartridges.
Emperor	72×48	......	......
Antiquarian	$53 \quad 31$	......	......
Double Imperial	......	......	44×30
Double Elephant	$40 \times 26\frac{3}{4}$	......	40×26
Atlas	33×26	......	......
Colombier	$34\frac{1}{2} \times 23\frac{1}{2}$	......	......
Imperial	34×22	30×22	30×22
Elephant	28×23	30×23	28×23
Super Royal	27×19	$27\frac{1}{2} \times 20\frac{1}{2}$	$27\frac{1}{2} \times 19\frac{1}{2}$
Cartridge or Log	......	......	26×21
Royal	24×19	25×20	25×20
Medium	$22 \times 17\frac{1}{2}$	24×19	......
Demy	$20 \times 15\frac{1}{2}$	$22\frac{1}{2} \times 17\frac{1}{2}$	$22 \times 17\frac{1}{2}$
Music Demy	......	$20\frac{3}{4} \times 14\frac{3}{4}$	......
Large Post	$21 \times 16\frac{1}{2}$	......	......
Copy	20×16	......	$20 \times 16\frac{1}{4}$
Post	$19 \times 15\frac{1}{4}$	$19\frac{1}{4} \times 15\frac{1}{4}$	......
Foolscap	$17 \times 13\frac{1}{2}$	$17 \times 13\frac{1}{2}$	......
Pott	$15 \times 12\frac{1}{2}$	......	......
Sheet and Half Pott	$22\frac{1}{2} \times 12\frac{1}{2}$	......	......
Sheet and Third Cap.	$22 \times 13\frac{1}{4}$	......	......
Sheet and Half Cap	$24\frac{1}{2} \times 13\frac{1}{2}$	......	......
Sheet and Half Post	......	$23\frac{1}{2} \times 19\frac{1}{2}$	......
Double Foolscap	$26\frac{1}{2} \times 16\frac{5}{8}$	27×17	......
Double Crown	.....	30×20	30×20
Double Post	......	$31\frac{1}{2} \times 19\frac{1}{2}$	......
Double Demy	.....	$35 \times 22\frac{1}{2}$	$35\frac{1}{2} \times 22\frac{1}{2}$

Relative Weights of a Ream containing 480, 500, or 516 Sheets.

Ream of 480 Sheets.	Ream of 500 Sheets.		Ream of 516 Sheets.		Ream of 480 Sheets.	Ream of 500 Sheets.		Ream of 516 Sheets.	
lb.	*lb.*	*oz.*	*lb.*	*oz.*	*lb.*	*lb.*	*oz.*	*lb.*	*oz.*
7	7	4	7	8	39	40	10	41	15
8	8	5	8	9	40	41	10	43	0
9	9	6	9	10	41	42	11	44	1
10	10	6	10	12	42	43	12	45	2
11	11	7	11	13	43	44	12	46	3
12	12	8	12	14	44	45	13	47	5
13	13	8	13	15	45	46	14	48	6
14	14	9	15	1	46	47	14	49	7
15	15	10	16	2	47	48	15	50	8
16	16	10	17	3	48	50	0	51	9
17	17	11	18	4	49	51	0	52	11
18	18	12	19	5	50	52	1	53	12
19	19	12	20	7	51	53	2	54	13
20	20	13	21	8	52	54	2	55	14
21	21	14	22	9	53	55	3	56	15
22	22	14	23	10	54	56	4	58	1
23	23	15	24	11	55	57	4	59	2
24	25	0	25	12	56	58	5	60	3
25	26	0	26	14	57	59	6	61	4
26	27	1	27	15	58	60	6	62	5
27	28	2	29	0	59	61	7	63	7
28	29	2	30	1	60	62	8	64	8
29	30	3	31	3	61	63	8	65	9
30	31	4	32	4	62	64	9	66	10
31	32	4	33	5	63	65	10	67	11
32	33	5	34	6	64	66	10	68	13
33	34	6	35	7	65	67	11	69	14
34	35	6	36	8	66	68	12	70	15
35	36	7	37	10	67	69	12	72	0
36	37	8	38	11	68	70	13	73	1
37	38	8	39	12	69	71	14	74	3
38	39	9	40	13	70	72	14	75	4

Equivalent Weights per Ream of Writing Paper of various sizes.

Foolscap, 16¾ × 13¼		Pinched Post, 18½ × 14¾		Post, 19 × 15¼		Large Post, 21 × 16½		Extra Large Post, 22½ × 17¾		Royal, 24 × 19¼		Super Royal, 27 × 19¼		Imperial, 30 × 22	
lb.	*oz.*	*lb.*	*oz.*	*lb.*	*oz.*	*lb.*	*oz.*	*lb.*	*oz.*	*lb.*	*oz.*	*lb.*	*oz.*	*lb.*	*oz.*
7	11	9	7	10	1	12	0	13	14	16	0	18	0	22	14
8	5	10	4	10	14	13	0	15	0	17	6	19	8	24	13
9	0	11	1	11	12	14	0	16	3	18	11	21	5	26	11
9	10	11	13	12	9	15	0	17	5	20	0	22	10	28	10
10	4	12	10	13	6	16	0	18	8	21	6	24	1	30	8
10	14	13	6	14	4	17	0	19	10	22	11	25	9	32	7
11	9	14	3	15	1	18	0	20	13	24	0	27	1	34	5
12	3	15	0	15	15	19	0	21	15	25	6	28	9	36	4
12	13	15	12	16	12	20	0	23	2	26	11	30	1	38	2
13	7	16	9	17	9	21	0	24	4	28	0	31	9	40	1
14	2	17	6	18	7	22	0	25	7	29	6	33	1	41	15
14	12	18	2	19	4	23	0	26	9	30	11	34	9	43	14
15	6	18	15	20	2	24	0	27	12	32	1	36	1	45	12
16	0	19	11	20	15	25	0	28	14	33	6	37	9	47	11
16	11	20	8	21	12	26	0	30	1	34	11	39	1	49	9
17	5	21	5	22	10	27	0	31	3	36	1	40	9	51	8
17	15	22	1	23	7	28	0	32	6	37	6	42	1	53	6
18	9	22	14	24	5	29	0	33	8	38	11	43	9	55	5
19	4	23	11	25	2	30	0	34	11	40	1	45	1	57	3
19	14	24	7	26	0	31	0	35	13	41	6	46	9	59	2
20	8	25	4	26	13	32	0	37	0	42	11	48	1	61	0
21	3	26	0	27	10	33	0	38	2	44	1	49	9	62	15
21	13	26	13	28	8	34	0	39	5	45	6	51	1	64	13
22	7	27	10	29	5	35	0	40	7	46	12	52	9	66	12
23	1	28	6	30	3	36	0	41	10	48	1	54	1	68	11
23	12	29	3	31	0	37	0	42	12	49	6	55	10	70	9
24	6	30	0	31	13	38	0	43	15	50	12	57	2	72	8
25	0	30	12	32	11	39	0	45	1	52	1	58	10	74	6
25	10	31	9	33	8	40	0	47	4	53	6	60	2	76	5

Sizes of Account Books.

Description.	Folio.				Quarto.				Octavo.			
	Broad.		Long.		Broad.		Long.		Broad.		Long.	
	Length.	Width.	Length.	Width.	Length.	Width.	Length.	Width.	Length.	Width.	Length.	Width.
	Ins.	Ins.	Ins.	Ins.	Ins.	Ins.	Ins.	Ins.	Ins.	Ins.	Ins.	Ins.
Foolscap ...	$12\frac{1}{2}$	8	$15\frac{1}{2}$	$6\frac{1}{4}$	$7\frac{1}{2}$	$6\frac{1}{4}$	$12\frac{1}{2}$	4	6	$3\frac{5}{16}$	$7\frac{1}{2}$	3
Demy	$14\frac{1}{4}$	$9\frac{1}{2}$	$18\frac{1}{2}$	$7\frac{3}{4}$	9	$7\frac{3}{4}$	$14\frac{1}{2}$	$4\frac{1}{2}$	7	$4\frac{1}{2}$	9	$3\frac{3}{8}$
Medium ...	$16\frac{1}{4}$	$10\frac{1}{4}$	$20\frac{1}{2}$	$8\frac{1}{4}$	$10\frac{1}{4}$	$8\frac{1}{4}$	$16\frac{1}{4}$	$5\frac{1}{8}$	8	5	$10\frac{1}{2}$	4
Royal	$18\frac{1}{2}$	$11\frac{1}{2}$	23	9	$11\frac{1}{2}$	9	$18\frac{1}{2}$	$5\frac{1}{2}$	9	$5\frac{1}{2}$	$11\frac{1}{2}$	$4\frac{3}{4}$
Super Royal	$18\frac{1}{2}$	13	26	$9\frac{1}{4}$	13	$9\frac{1}{4}$	$18\frac{1}{2}$	$6\frac{1}{2}$	9	$6\frac{1}{2}$	13	$4\frac{1}{2}$
Imperial ...	$20\frac{1}{2}$	$14\frac{1}{2}$	29	$10\frac{1}{4}$	$14\frac{1}{2}$	$10\frac{1}{4}$	$20\frac{1}{2}$	$7\frac{1}{2}$	10	$7\frac{1}{4}$	$14\frac{1}{4}$	5

Sizes of Glazed Boards.

	Inches.
Foolscap	$17\frac{1}{2} \times 11\frac{1}{2}$
Demy	22×18
Royal	24×19
Royal Extra	$25\frac{1}{2} \times 20$
Double Foolscap........................	29×18
Super Royal	$29 \times 21\frac{1}{2}$
Imperial	31×23

Sizes of Brown Papers.

	Inches.
Casing	46×36
Double Imperial	44×29
Elephant	34×24
Double Four Pound	31×22
Imperial Cap	29×22
Haven Cap	26×21
Bag Cap	$24 \times 19\frac{1}{2}$
Kent Cap	21×18

Sizes of Letter, Note Paper, &c.

	Inches.
Medium 4to	10¾ × 8⅜
Medium 8vo	8⅜ × 5⅜
Demy 4to	9⅜ × 7⅜
Demy 8vo	7¼ × 4¼
Demy 16mo	4⅝ × 3⅝
Large Post 4to	10 × 8
Large Post 8vo	8 × 5
Post 4to	9 × 7⅜
Post 8vo	7⅛ × 4½
Post 16mo	4⅜ × 3⅝
Copy 4to	9⅝ × 7¼
Copy 8vo	7¼ × 4⅝
Foolscap 4to	8 × 6⅜
Albert	6 × 3⅞
Queen	5⅜ × 3½
Prince of Wales	4½ × 3
Foolscap	12¼ × 8

Sizes of Sugar Papers.

	Inches.
Double Two Pound	24 × 16
Large ditto	27 × 17
Double Small Hand	30 × 19
Royal Hand	25 × 20
Lumber Hand	23½ × 18
Middle Hand	22½ × 16
Purple Copy Loaf	22½ × 16½
Ditto Double ditto	23 × 16½
Ditto Powder ditto	26 × 18½
Ditto Single ditto	28 × 22
Ditto Elephant	29 × 24
Purple Lump Loaf	33 × 23
Ditto Titler	35 × 20

Sizes of Cards.

Inches.

Large	$4\frac{1}{2}$	$\times$ 3
Carte de Visite	$4\frac{1}{8}$	$\times$ $2\frac{1}{2}$
Small	$3\frac{1}{2}$	$\times$ $2\frac{1}{2}$
Reduced Small	$3\frac{1}{2}$	$\times$ $2\frac{1}{8}$
Extra Thirds	3	$\times$ $1\frac{7}{8}$
Thirds	3	$\times$ $1\frac{1}{2}$
Half Small	$2\frac{1}{2}$	$\times$ $1\frac{3}{4}$
Town Size	3	$\times$ 2
Half Large	3	$\times$ $2\frac{1}{4}$
Double Small	5	$\times$ $3\frac{1}{2}$
Double Large	6	$\times$ $4\frac{1}{2}$
Quadruple Small	7	$\times$ 5
Quadruple Large	9	$\times$ 6

Sizes of French Printing Papers with English Equivalents.

	Inches.	Centimètres.
Pot	12.2×16.5	31×42
Poulet	8.6×11.0	22.5×28
Couronne	14.2×18.1	36×46
Ecu	15.7×20.4	40×52
Coquille	17.7×22.0	45×56
Cloche-Normande	13.8×20.4	35×52
Tellière	13.0×17.3	33×44
Griffon	13.8×17.7	35×45
Petit Raisin	12.6×17.0	32×43
Carré	17.7×22.0	45×56
Cavalier	18.5×23.6	47×60
Raisin, *or* Grand Raisin	19.7×25.6	50×65
Petit Jésus	21.6×27.5	55×70
Jésus-Musique	22.0×27.5	54×70
Grand Jésus	22.4×29.5	56×75
Petit Colombier, *or* Soleil	23.6×31.5	60×80
Grand Colombier	24.4×35.4	62×90
Grand Aigle	28.8×40.0	73×102
Grand Monde	35.4×47.1	90×120

Sizes of English Printing Papers with French Equivalents.

	Inches.		Centimètres.
Pott	12.5 × 15.5	...	31.7 × 39.4
Foolscap	13.5 × 17	...	34.3 × 43.2
Post	15.7 × 19.5	...	40.0 × 49.6
Crown	15 × 20	...	38.1 × 50.8
Demy	17.5 × 22.5	...	44.4 × 59.1
Medium	18.5 × 23.5	...	47.6 × 59.7
Royal	20 × 25	...	50.8 × 62.5
Super Royal	20.5 × 27.5	...	52.0 × 69.9
Imperial	22 × 30	...	55.9 × 76.2
Double Pott	15.5 × 25	...	39.4 × 62.5
Double Foolscap	17 × 27	...	43.2 × 68.6
Double Post	19.5 × 31.5	...	49.6 × 40.0
Double Crown	20 × 30	...	50.8 × 76.2
Double Demy	22.5 × 35	...	57.1 × 88.8
Double Royal	25 × 40	...	62.5 × 101.6
Double Super Royal	27.5 × 41	...	69.9 × 104

Sizes of German Papers with French Equivalents.

Writing Paper.

	Inches.		Centimètres.
Schlängle	12. 2 × 15.3	...	31 × 39.5
Canzlei	12. 9 × 16.5	...	33 × 42
,, (Untrimmed)	... 13.10 × 16.9	...	34 × 43
Propatria	14. 1 × 17.3	...	36 × 44.5
Löwen	14. 3 × 19.2	...	37 × 48

Paper for Account Books and for Drawings.

	Inches.		Centimètres.
Klein Median	15.6 × 20	...	40 × 51
Median	16.5 × 21.5	...	42 × 54
Gross Median	17.3 × 23.2	...	44 × 58
Klein Royal	19.2 × 24.8	...	48 × 63
Noten Royal (for Music)	19.9 × 26.3	...	50 × 67
Gross Royal	20.7 × 27.1	...	52 × 68
Super Royal	21.5 × 28.2	...	54 × 72
Imperial	22.8 × 29.8	...	58 × 76
Klein Adler	24.4 × 35.5	...	62 × 90
Elephant	26.3 × 36.3	...	67 × 92

Sizes of Italian Papers with French Equivalents.

	Inches.			Centimètres.	
Ottavina	5.3	×	8.5 ...	13.5 ×	21
Sestina	7.1	×	9.0 ...	18 ×	22.5
Quartina	8.3	×	10.6 ...	21 ×	27
Mezzanella	9.0	×	14.2 ...	23 ×	36
Olandina	9.8	×	15.3 ...	25 ×	39
Quadrotta (formato francesé)	10.4	×	16.5 ...	26.5 ×	42
., (formato Italiano)	10.6	×	17.4 ...	27.5 ×	44.5
„ (formato tedesco)	11.4	×	18.9 ...	29 ×	48
Processo, *or* Notarile	10.2	×	15.0 ...	26 ×	38
Protocollo, *or* Pellegrina	12.2	×	16.5 ...	31 ×	42
Rispetto	13.0	×	17.7 ...	33 ×	45
Stato, *or* Leona	14.1	×	18.9 ...	36 ×	48
Bastarda	16.5	×	22.1 ..:	42 ×	56
Realino, *or* Mezzana	17.7	×	23.6 ..:	45 ×	60
Reale	19.7	×	25.5 ...	50 ×	65
Realone	20.4	×	27.1 ...	52 ×	69
Imperialino	21.2	×	29.9 ...	54 ×	76
Imperiale	24.0	×	32.0 ...	61 ×	81
Elefante	26.0	×	37.8 ...	66 ×	96
Aquila	27.5	×	39.4 ...	70 ×	100

Crystalline Coating for Paper.—This may be obtained, it is stated, by mixing a very concentrated cold solution of salt with dextrine and laying the thinnest coating of the fluid on the surface to be covered by means of a broad, soft brush. After drying, the surface has a beautiful, bright, mother-of-pearl coating, which, in consequence of the dextrine, adheres firmly to paper and wood. The coating may be made adhesive to glass by doing it over with an alcoholic shellac solution. Sulphate of magnesia, acetate of soda and sulphate of tin are among the salts which produce the most attractive crystalline coatings. Paper must first be sized, otherwise it will absorb the liquid and prevent the formation of crystals.

Table showing Cost of Paper per 100.

Price per Ream.	5/-		6/-		7/-		8/-		9/-		10/-		11/-		12/-		13/-	
Sizes.	*s.*	*d.*	*s.*	*d.*	*s.*	*d.*	*s.*	*d.*	*s.*	*d.*	*s.*	*d.*	*s.*	*d.*	*s.*	*d.*	*s.*	*d.*
16mo.	0	0¾	0	1	0	1¼	0	1¼	0	1½	0	1¾	0	1¾	0	2	0	2
12mo.	0	1¼	0	1¼	0	1½	0	1¾	0	2	0	2¼	0	2½	0	2¾	0	2¾
8vo.	0	1¾	0	2	0	2¼	0	2½	0	3	0	3¼	0	3½	0	4	0	4¼
4to.	0	3½	0	4	0	4½	0	5¼	0	6	0	6½	0	7¼	0	7¾	0	8½
Third	0	4½	0	5¼	0	6	0	7	0	8	0	8¾	0	9½	0	10½	0	11¼
Half-sheet	0	7¼	0	8	0	9	0	10½	0	11¾	1	1	1	2¼	1	3½	1	5
Sheet	1	1	1	3½	1	6	1	9	1	11¼	2	2	2	4½	2	7¼	2	10

Price per Ream.	14/-		15/-		16/-		17/-		18/-		19/-		20/-		21/-		24/-	
Sizes.	*s.*	*d.*	*s.*	*d.*	*s.*	*d.*	*s.*	*d.*	*s.*	*d.*	*s.*	*d.*	*s.*	*d.*	*s.*	*d.*	*s.*	*d.*
16mo.	0	2¼	0	2½	0	2¾	0	2¾	0	3	0	3¼	0	3¼	0	3½	0	4
12mo.	0	3	0	3¼	0	3½	0	3¾	0	4	0	4¼	0	4½	0	4½	0	5½
8vo.	0	4½	0	5	0	5¼	0	5½	0	6	0	6¼	0	6½	0	7	0	8
4to.	0	9	0	9¾	0	10½	0	11	0	11¾	1	0¼	1	1	1	1½	1	3½
Third	1	0¼	1	1	1	2	1	2¾	1	3¾	1	4½	1	5¼	1	6	1	9
Half-sheet	1	6¼	1	7½	1	9	1	10	1	11½	2	0¾	2	2	2	3¼	2	7
Sheet	3	0½	3	3	3	5½	3	8	3	11	4	1½	4	4	4	6½	5	2½

Table showing Cost of Paper per 1,000.

Price per Ream.	5/-		6/-		7/-		8/-		9/-		10/-		11/-		12/-		13/-	
Sizes.	*s.*	*d.*	*s.*	*d.*	*s.*	*d.*	*s.*	*d.*	*s.*	*d.*	*s.*	*d.*	*s.*	*d.*	*s.*	*d.*	*s.*	*d.*
16mo.	—		0	9¾	0	11½	1	1	1	2¾	1	4¼	1	6	1	7½	1	9
12mo.	0	11	1	1	1	3	1	5½	1	7½	1	9¾	2	0	2	2	2	4
8vo.	1	4¼	1	7½	1	10¾	2	2	2	5¼	2	8½	2	11¾	3	3	3	6¼
4to.	2	8½	3	3	3	9½	4	4	4	10½	5	5	5	11½	6	6	7	0½
Third	3	7½	4	4	5	1	5	9½	6	6	7	3	7	11½	8	8	9	5
Half-sheet	5	5	6	6	7	7	8	8	9	9	10	10	11	11	13	0	14	1
Sheet	10	10	13	0	15	2	17	4	19	6	21	8	23	10	26	0	28	2

Price per Ream.	14/-		15/-		16/-		17/-		18/-		19/-		20/-		21/-		24/-	
Sizes.	*s.*	*d.*	*s.*	*d.*	*s.*	*d.*	*s.*	*d.*	*s.*	*d.*	*s.*	*d.*	*s.*	*d.*	*s.*	*d.*	*s.*	*d.*
16mo.	1	10¾	2	0½	2	2	2	3½	2	5¼	2	7	2	8½	2	10	3	3
12mo.	2	6¼	2	8½	2	10¾	3	1	3	3	3	5	3	7¼	3	9½	4	4
8vo.	3	9½	4	0¾	4	4	4	7¼	4	10½	5	1¾	5	5	5	8½	6	6
4to.	7	7	8	1½	8	3	9	2½	9	9	10	3½	10	10	11	4½	13	0
Third	10	2	10	10	11	7	12	3½	13	0	13	9	14	5½	15	2	17	4
Half-sheet	15	2	16	3	17	4	18	5	19	6	20	7	21	8	22	9	26	0
Sheet	30	4	32	6	34	8	36	10	39	0	41	2	43	4	45	6	52	0

Sizes of Printing Paper, subdivided.

	Broadside.	Folio.	Quarto.	Octavo.	16mo.	32mo.
D. Sup. Royal	27½×41	20½×27½	13¾×20½	10¼×13¾	6¾×10¼	5×6¾
Double Royal	25×40	20×25	12½×20	10×12½	6¼×10	5×6¼
Double Demy ...	22½×35	17½×22½	11¼×17½	8¾×11¼	5½×8¾	4½×5½
Double Crown	20×30	15×20	10×15	7½×10	5×7½	3¾×5
Double Post ...	19½×31½	15¾×19½	9¾×15¾	8×9¾	5×7½	4×5
Double Fcap. ...	17×27	13½×17	8½×13½	6¾×8½	4¾×6¾	4¾×8⅜
Double Pott ...	15½×25	12½×15½	7¾×12½	6¼×7¾	4×6¼	3×4
Imperial	22×30	15×22	11×15	7½×11	5½×7½	3¾×5½
Super Royal ...	20½×27½	13¾×20½	10¼×13¾	7×10¼	5×7	5×8⅜
Royal	20×25	12½×20	10×12½	6¼×10	5×6¼	3×5
Medium	18½×23½	11¾×18½	9¼×11¾	6×9¼	4½×6	3×4½
Demy	17½×22½	11¼×17½	8¾×11¼	5½×8¾	4¼×5½	2¾×4¼
Crown	15×20	10×15	7½×10	5×7½	3¾×5	2½×3¾
Post	15¾×19½	9¾×15¾	8×9¾	5×8	4×5	2½×4
Foolscap	13½×17	8½×13½	6¾×8½	4¼×6¾	3½×4¾	2×3½
Pott	12½×15½	7¾×12½	6¼×7¾	4×6¼	3×4	2×3

To Make Emery Paper.—Fix a sheet of stout blotting paper on a board, gluing it round the edge. Put emery powder into a sifter, the mesh of which has the requisite degree of fineness, and, rapidly covering the paper with thin hot glue, shake the sifter lightly over the paper until it is evenly covered, and leave to cool. When dry, detach the paper and shake it vigorously to remove loose grains.

Table for Giving out Paper.

Number of Copies	1		2		3			4			8			12			16			Number of Copies
	Q.	S.	Q.	S.	Q.	S.	O.	Q.	S.	O.	Q.	S.	O.	Q.	S.	O.	Q.	S.	O.	
50	2	2	1	1	0	17	1	0	13	2	0	7	6	0	5	10	0	4	14	50
100	4	4	2	2	1	10	2	1	1	0	0	13	4	0	9	8	0	7	12	100
150	6	6	3	3	2	2	0	1	14	2	0	19	2	0	13	6	0	10	10	150
200	8	8	4	4	2	19	1	2	2	0	1	1	0	0	17	4	0	13	8	200
250	10	10	5	5	3	12	2	2	15	2	1	8	6	0	21	2	0	16	6	250
300	12	12	6	6	4	4	0	3	3	0	1	14	4	1	1	0	0	19	4	300
400	16	16	8	8	5	14	2	4	4	0	2	2	0	1	10	8	1	1	0	400
500	20	20	10	10	6	23	1	5	5	0	2	15	4	1	18	4	1	8	12	500
750	31	6	15	15	10	10	0	7	20	2	3	22	2	2	15	6	1	23	2	750
1000	41	16	20	20	13	22	2	10	10	0	5	5	0	3	12	8	2	15	8	1000
1500	62	12	31	6	20	20	0	15	15	0	7	20	4	5	5	0	3	22	4	1500
2000	83	8	41	16	27	19	1	20	20	0	10	10	0	6	23	4	5	5	0	2000
3000	125	0	62	12	41	16	0	31	6	0	15	15	0	10	10	0	7	20	8	3000
4000	166	16	83	8	55	14	2	41	16	0	20	20	0	13	22	8	10	10	0	4000
5000	208	8	104	4	69	11	1	52	2	0	26	1	0	17	9	4	13	1	8	5000

Note.—Q *means quires,* s *sheets, and* O *the overplus copies.*

Carbon Paper.—To make carbon paper take of clear lard, five ounces; beeswax, one ounce; Canada balsam, one-tenth ounce; lampblack, q. s. Melt by aid of heat, and mix. Apply with a flannel dauber, removing as much as possible with clean woollen rags.

Impermeable Wrapping Paper.—Dissolve one and a half pounds of soap in a quart of water; then dissolve two ounces of gum arabic and six ounces of glue in another quart of water. Mix the two solutions; warm the mixture; dip the paper in the liquid; pass it between two rolls (a clothes-wringer, for example), and put it to dry. In default of rolls, hang the paper up, that it may drip well, or better, pass it between two sheets of dry paper. Then let dry in mild temperature.

To Soften Harsh Papers.—Almost any kind of paper, however harsh in texture, may be rendered soft and flexible by heating it in a solution of acetate of soda, or of potash dissolved in four to ten times its weight of water. For permanent paper to twenty parts of this solution one part of starch or dextrine is added. If the paper has to be made transparent, a little of a solution containing one part soluble glass in four to eight parts water is added. To render the paper fit for copying without being made wet, to the acetate solution chromic acid or ferro-cyanide of potassium is added.

Satin Paper.—The Belgian or satin paper, which has the appearance of silk or satin, has a calendered and sized book paper as a foundation layer. The paper is printed with the zinc white ground in No. 3 varnish, and when dry the sheets are run through a calendering machine. Where the latter cannot be had, take a lithographic stone polished as smoothly as possible, and with oxalic acid, water, and paper, make a paste (oxalic acid must be powdered before the water is poured on it), and rub it over the stone until it has the appearance of a looking-glass. This can best be done by using a large piece of cork smooth on the bottom, and a piece of woollen cloth or flannel over it. With this dabber rub the oxalic acid on the stone with heavy pressure in the same manner as in stone grinding. When the printed and asbestos-dusted sheets are pulled through the press—the printed side, of course, to the polished surface of the stone—the asbestos will be by this pressure fastened to the sheets by its lengthy fibres, and give thereby the satin-like appearance of which we have spoken.

Tracing Cloth is thin muslin sized with isinglass and passed through polished rolls heated by steam. Tracing paper is either sized with isinglass and calendered, or oiled with linseed oil.

To Detect Arsenic in Papers impregnate the paper with a solution of nitre, dry, and burn it on a plate; the ashes are boiled with dilute caustic potash, the liquor filtered, and sulphuric acid and permanganate solution added, until no more is decolorized; finally, the cold filtered liquid is treated in a flask with zinc and sulphuric acid, two papers being placed on the neck, one being impregnated with lead acetate, the other with silver nitrate ; unless sulphuretted hydrogen be formed, the first is not blackened, whilst the second is blackened if arseniuretted hydrogen be evolved, *i.e.*, if arsenic be present in the paper.

Papers Sized with Resin Size are found to have a more or less acid reaction due to free sulphuric acid, which has never been observed in samples sized with animal glue. The acid is probably derived from the alum or aluminum sulphate used in sizing, which is decomposed by contact with the vegetable fibre, as it takes place in dyeing, a basic salt being deposited upon the fibre, and a portion of acid liberated.

To Wet Down Paper.—Place a wetting-board near the trough ; on it lay half-a-dozen dry sheets of the paper to be wetted. Then take half-a-dozen other sheets and pass them together through the water ; then lay half-a-dozen sheets of dry paper on these, and go on in the same manner till you have finished. If the paper was delivered folded, it must be opened out, and you must be careful to turn the backs or ridges sometimes up and sometimes down, in order that it may eventually come out flat. When the wetting process is finished, place a board on the top of the heap, and place weights on it—a small weight at first—say twenty pounds, and after an hour increase it to sixty, and then put on as much pressure as you like. Leave the heap for twelve hours. If it opens out too wet, intersperse some dry sheets throughout the heap, and squeeze it for some time in a screw or hydraulic press ; if too dry, sprinkle every dozen or twenty sheets and press as before.

For Paper Sizing, caseine of milk, which has the same chemical composition as egg albumen, is a good and cheap substitute. It may be dissolved in water that is slightly alkaline, especially in very dilute aqua ammonia.

To Prepare Paper for Transparent Printing.—

The secret of this lies entirely in the preparation of the paper. Thin plate paper is coated with good flour or starch paste evenly laid on with a brush and dried slowly before a fire. When dry another coating, this time of a strong solution of gum or gum and starch, is given, and when this is dry the sheet is run once or twice through a rolling press, or pressed between zinc plates in a hydraulic press, which ensures an even surface for printing on. If the tablet is to be affixed to glass the printing is done in the ordinary rotation of colours in the usual way. The coating of gum rendering the surface adhesive a slight damping is all that is necessary to affix the tablet to the glass.

Waterproof Labels.—

The label having been properly pasted on and dried, coat it first with a size prepared by adding as much alcohol to thick mucilage or gum arabic as this will stand without precipitating. When the sizing is dry, the label is brushed over with a solution of 50 parts of mastic and 1 part of storax in 165 parts of alcohol. In making the latter solution, it is recommended to add about 25 parts of sand, which mixes with the mastic and permits the alcohol to penetrate the mass more rapidly.

Waterproof Paper.—

The following is given as a method of rendering paper impervious to water:—To a weak solution of common glue add acetic acid—about five per cent. Then dissolve in distilled water about seven per cent. of bichromate of potassium. Mix the solutions together, pass the sheets through the compound, and then hang up the paper to dry. When dry, polish and surface in the usual way.

Enamel Paper.—

The pigments used in enamelling paper are metallic substances such as will spread smoothly and take a polish, and include white lead, oxide of zinc, sulphate of barytes, china clay, whiting, chalk, in a menstruum, or upon a previous coating of glycerine, size, collodion, water, varnish, etc., polished between calendering or burnishing cylinders.

Fire and Water-Resisting Paper.

—It is said paper can be prepared so as to resist the effects of both fire and water by treating it with a solution of asbestos, common salt, and alum. The paper is plunged in this mixture and afterwards dried. It is then treated with a gum dissolved in alcohol, and afterwards dried and surfaced between rollers of metal. The solution strengthens the paper and renders it fireproof; the gum makes it impermeable to water.

To Make Blue Sugar Paper.

—A very good recipe for this is as follows:—To 1½ lb. of pulp add 24 oz. verdigris, 3 lb. alum, and 50 to 55 cans water, which have to be boiled together. After the pulp has lain for a few days it is taken out, the water drained off, and the pulp placed in a decoction of Brazil wood, being then worked up as for ordinary paper.

Anti-Forgery Paper.

—To prevent alterations in writing, the following process of preparing paper is recommended by an American inventor:—Add to the sizing five per cent. of cyanide of potassium and sulphide of antimony, and run the sized paper through a thin solution of sulphate of manganese or copper. Any writing on this paper with ink made from nut-galls and sulphate of iron can neither be removed with acids nor erased mechanically. Any acid will immediately change the writing from black to blue or red. Any alkali will change the paper to brown. Any erasure will remove the layer of colour and expose the white ground of the paper, since the colour of the paper is only fixed to the outside of the paper without penetrating it.

Uninflammable Paper.

—The following is given as a cheap mode of rendering fabrics and paper uninflammable : Four parts of borax and three parts sulphate of magnesia are shaken up together just before being required. The mixture is then dissolved in from twenty to thirty parts of warm water. Into the resulting solution the articles to be protected from fire are immersed, and when they are thoroughly soaked, they are dried, preferably in the open air.

M

A New Process for Watermarking Paper consists in transferring to the dandy roll a design in relief, previously executed on a sheet of paper with a small tube-pen containing a specially prepared enamel. The design, as drawn, is placed on the roll or mould and after some hours the paper is removed by damping, leaving the enamel design fixed on the wire-cloth. With the tube-pen the lines in relief are then added, imitating the wire-mark, and covered with varnish, producing a filigrane. When done with, this watermark can be removed without damaging the roll or mould, and a fresh design applied, and in case of accident it can be easily repaired.

Coloured Wall-Papers.—In the manufacture of coloured wall-papers the white paper is first coated with a uniform "ground" tint. This is done by machine. The paper cannot be rolled until perfectly dry. A very high temperature and a considerable number of persons are necessary for that purpose, and the losses during drying, both accidental and by the negligence of employés, are considerable. As it is necessary to have on hand a great variety of ground tints, this operation requires a large stock of coloured papers. The designs for the coloured figures are reproduced by engravers, each colour on a separate wooden cylinder. The colours used for printing are generally in the form of paste mixed with glue. The engraving must therefore have enough relief for the purpose of printing, to prevent the details of the colours from blending.

How Hand-made Papers are Made.—Hand-made paper is formed, one sheet at a time, on what is technically termed a mould—a light wooden frame traversed with thin slats $1\frac{1}{4}$ inches apart, and the whole covered with fine wire-cloth ; it is fitted loosely inside of another frame, which is termed the "deckle," and forms an elevated border all around the mould. Machine paper is made on a continuous web of felt in a Fourdrinier machine. The difference between the two is that the paper made on a hand-mould is thoroughly felted in all directions and is subjected to no strain while being finished, and therefore retains all its normal strength in length and width. With machine-made papers the fibres lie more in one direction than in another, and in transferring from one section of the machine to another, and subsequently, while in the web form, it is strained in the same direction—a weakening process as compared with hand-made. For this reason manufacturers of ledger and similar papers in which strength is required are careful to keep the tension of the paper between the different sections of the machine as light as possible, but even then considerable difficulty is experienced in securing strength equal to that obtained by the hand process.

Straw Board is made altogether, or principally, from wheat or rye straw. The first process consists in boiling the straw with quicklime in a wooden digester, taking steam from a boiler. The straw is packed in layers with lime between, and the whole boiled for ten or twelve hours, according to circumstances. Straw is composed of a tube of woody fibre and cellular tissue, its surface containing silicates of potassa and soda with free silica. The woody fibre also contains silica. To this silica the straw of grain or grass owes in great part its strength. In boiling, the lime and the silica combine, leaving the straw in a soft pulpy state. The mass is now ground into pulp, and then drawn into a vat containing water, kept constantly agitated by a series of revolving arms. A wire-gauze cylinder is adjusted that revolves partially beneath the surface of the fluid mass. The pulp adheres to the gauze, and is carried to another cylinder, around which an endless belt of felt runs. The latter cylinder presses upon the gauze, and causes the pulp to adhere to the felt and condense so as to give it enough consistency to be taken up by yet another cylinder, called the "forming cylinder," one of a pair made of polished metal, and by these the pulp is strongly compressed. The pulp is wound round the "former" until the proper thickness, determined by an indicator, is obtained. Along the forming "cylinder" there is a groove planed out, through which the operator now rapidly passes a wooden knife, thus severing the soft board, and at the same time he unwinds the sheet and removes it. These sheets are then dried, completing the process. Woollen rags are sometimes ground and mixed with the straw pulp, making a much darker coloured and heavier board, and worth considerably more than the pure straw board. The white lining of these boards was pasted by hand, but now a machine does this at the rate of twenty sheets a minute. This machine is auto-matic; it pastes the boards, lays on the white paper from a continuous roll, dries, presses, and calenders them, so that when the boards leave the machine they are ready for the box manufacturer.

Washable Paper.—Writing and drawing paper first receive a thin coating of a mixture of glue, or some other suitable adhesive substance, with zinc white, chalk, barytes, etc., and the colour for producing the desired tint. They are then coated with cilicate of soda, to which a small quantity of magnesia has been added, and dried at a temperature of 25° C. during ten days or so. Paper thus treated is said to possess the property of preserving writing or drawing in lead pencil, chalk, or Indian ink.

Papier-Maché is made either by cementing together sheets of paper and afterwards coating them with oil, and then baking them, finishing with varnish, or it is produced by re-pulping old paper or from new pulp. In the latter case, wood is frequently employed, besides other fibrous material. To this end it is pulped in the same manner as for paper, and the pulp then compounded with the necessary constituents for its production, such as oil, pitch, resin, soda, sugar of lead, glue, etc. From this, in a pasty or doughy state, the articles are made by stamping or moulding, and then baking. Of course inlaying and such work is done while in a plastic state, or the papier-maché is made into blocks or sheets, then finished, and afterwards cut and worked up in the same manner as wood. The uses to which this substance is put are very numerous, as, for instance, the making of furniture, book-backs, albums, buttons, medals, picture-frames, and a host of other purposes.

Pinched Post.—It has been said that this size resulted from the double invention of paper-making machinery and paper-cutting machinery, the former overcoming the waste of the unfinished deckle edge of paper made by hand, and the latter saving every second shaving which the plough-cutting press necessitated. We still take our sizes, however, from the hand-mould deckle, and call post 19 × 15¼, and large post 21 × 16½. Country travellers representing some of the first houses in the paper trade show samples accordingly, although the reams when cut up and re-tied measure no more than 9 × 14. Indeed, it is useless to send out post octavo larger than 4½ × 7 if it is to be used with the standard sizes of envelopes. The graduation of the four leading sizes of writing-paper, post octavo, large post octavo, post quarto, and large post quarto, would be greatly improved by shortening the folio of both post and large post one inch. If such a change could be made with the general approval and consent of the trade. 14¼ × 19 and 15¼ × 21 would have quite sufficient margin for the present width of the page of both post and large post octavo to be maintained, and the depth of each page to be half as much again as its width, thus insuring greater symmetry than now exists. The proposed alteration would be as follows :—Post folio, 14¼ × 19; post quarto, 6¾ × 9; post octavo, 4¼ × 6¾. Large post folio, 15¼ × 21 ; large post quarto, 7½ × 10; large post octavo, 5 × 7½.

Tracing Paper is very expensive in the market, and everyone who would like to make his own may do it in the following manner :—Mix well together 75 parts of olive oil and 25 parts of benzine. With a brush put it on best tissue paper, and hang it up for about thirty-six hours. This is a very transparent tracing paper : until the benzine has evaporated it is also extremely inflammable.

Another Method.—Steep sheets of paper in a strong solution of gum arabic, and afterwards press each sheet between two dry sheets of similar paper, to take off the superfluity of the liquid. This will convert these sheets of paper into a tracing paper. The solution must be strong—about the consistency of boiled oil.

Preservative Transfer Paper.—In two quarts of water dissolve three decagrams dextrine, soak for about two days, then boil it and allow it to cool. Add six decagrams of gelatine, put over fire again and dissolve the gelatine, taking care that the substance does not get to boiling, for in this the gelatine, together with the dextrine, would lose some of its adhesiveness. Let the whole become of a molasses-like consistency. This solution, in a warm condition, is to be filtered, and with it a heavy supersized and calendered book paper may be coated like transfer paper. It is advisable to add to the solution about half a pint of alcohol and ten drops of carbolic acid, and two or three ounces of glycerine to avoid a curling up of the edges of the paper. This may vary with the state of the temperature and the condition of the atmosphere; where in dry days three ounces will be required, two ounces will suffice on wet days.

Rapid Drying of Printed Work.—It is frequently important that circulars should be sent out immediately after having been printed—an object that may at times be effected by using a well-glazed printing paper instead of writing paper. The former will absorb the ink so fast that, unless piled up more than is customary, it may be folded and cut at once. Common qualities of writing paper have a strong tendency to absorb ink; but the hard, highly glazed papers absorb ink so very slowly that weeks may elapse before it becomes dry. It is apparent, therefore, that to secure fast drying the printing must be done with as little ink as will suffice to do the job properly, or something else must be done to obviate a set-off. One method to accomplish this is the dusting of the freshly-printed sheets with some fine powder which will not soil the paper but will stick to the ink. For the paper a white powder is the best, so long as it is not used in sufficient quantity to obscure the ink. Powdered French chalk is in favour for this purpose, but it causes the paper to slip about during the process of folding and cutting, besides imparting to it a slippery feel easily discernible to the touch. Another drawback is that the pulverized French chalk being of unequal granulation, the coarsest particles are left to the last, and become plainly visible. Calcined magnesia has none of these undesirable qualities. It costs more in the first place than French chalk, but on account of its lightness—a much larger quantity going to the pound—is cheaper in the end. It does not render the paper slippery, is of uniform fineness, leaving no coarse particles as a residuum. It has, however, a tendency to stick together and remain among the sheets, but this can be prevented by attention. In using either of these or any other powders particular care must be taken not to dust them on the top of the printed heap, as that will increase the setting-off. The proper manner is to lay out a quantity of the powder in a convenient place, then take a pad of cotton or wool, or else a piece of loose cotton cloth, and dust the powder into that for a beginning; then take the printed sheet from the pile, lay it down in a clean place, and rub the powder over the ink, taking care to leave no superfluous quantities, pass it on as one of a new heap, and so continue the operation. Powder must be renewed on the pad or cloth about every other sheet, this varying with the size of the job. It is, however, always important not to use too much powder, or the over-dusted sheets will have to be handled over again to remove the superfluity.

To Prepare Paper for Negative Printing.—The best method of preparing, sensitizing, and fixing ordinary plain paper for printing from a negative in black on a dead white ground, is to dip the paper into a solution of ammonium chloride, then float on a silver bath, which will form silver chloride.

Parchment Paper may be rendered impervious to oils by steeping in a hot solution of gelatine, to which $2\frac{1}{2}$ or 3 per cent. of glycerine has been added, and drying. To render the same paper waterproof, it is soaked in sulphuret of carbon, containing in the solution 1 per cent. of linseed oil and 4 per cent. of caoutchouc.

Unsized Plate Paper may be made impervious to moisture by immersing it in a solution of mastic in oil of turpentine, afterwards drying by a gentle heat.

How to Test Wood-Paper.—Lack of durability in wood-paper is oftentimes an embarrassment to book publishers, who cannot always decide upon the material which goes to compose the paper submitted to them for examination. The most certain way to discover wood-pulp paper is to take a drop of an acid composed of three-fourths nitric acid and one-fourth sulphuric acid. The paper wetted with this one drop will have a brown stain if there is wood-pulp in the paper. Experiments recently made in the testing of paper by means of this acid showed the following results:—White paper, entirely free from wood-pulp, is barely coloured by the acid, the part wetted taking on a slight grey tint after drying. Wood-pulp paper assumes a dark brown colour immediately on the application of the acid. With a very little experience the amount of wood in the paper under test may be approximated from the rapidity of the discoloration, the shade of the stain, and the dimensions of the grey-violet ring around the spot produced by the acid. In coloured papers the changes worked by the acid vary: blue wood-pulp paper gives a green stain, red paper a yellowish brown, green paper a reddish brown.

Weights of Different Sizes of Paper.—It often happens that a printer having a paper of a certain size and quality with a known weight per ream, wants to know what will be the weight of a ream of the same kind of paper in another size. The following example will show the means of ascertaining this:—

Example : I have a 24 lb. demy paper: what will be the weight of a ream of the same paper in double crown size?

The size of a sheet of demy is $17\frac{1}{2} \times 22\frac{1}{2}$ inches, and that of a sheet of double crown is 20×30 inches. To ascertain the weight of the ream of double crown, multiply 24 lbs. by (20×30) and divide by $(17\frac{1}{2} \times 22\frac{1}{2})$, i.e., 24×600 divided by $393\frac{1}{4}$ equals $36\frac{1}{4}$ lb. *Ans.*

The Rule is : Multiply the weight of the paper you have by the size (in square inches) of a sheet of the paper whose weight you require to know, and divide by the size (in square inches) of a sheet of the first-named paper

To Clean Glazed Boards.—These boards soiled with ink may be cleaned with a little turpentine rubbed on with a piece of flannel and finished with a soft duster.

Oiled Sheets for Letter-Copying.—The only preparation for this paper is to brush them over on both sides with boiled oil mixed with litharge. The sheets must be hung up to dry singly, as if placed in a heap too soon they are apt to generate heat.

On the Right and Wrong Sides of Paper.—Roughness of surface cannot be said to invariably indicate the wrong side of paper. Some misconception prevails on this point : but proof is readily obtained from papers for crayon and chalk drawings. The roughest there is the right side. On opening a ream of flat paper (*i.e.*, unfolded), the right side is the top side. When paper is folded into quires it is right side out. The lettering of the watermarks can only be read from the right side of the paper. When papers are azure laid, yellow wove, or blue, they are, if machine-made, usually darker on the wrong side ; if hand-made, the right side is the darker. Some of these characteristics may be absent, and then an independent test becomes necessary. This is found in the wire cloth mark. When everything else fails this points out the wrong side. There are but few exceptions to this rule, since it is seldom the exhaust boxes act so lightly that the finishing obliterates the marks they leave. The wrong side of a granite paper is denoted by all the fibre being set in the same direction. It is worth mentioning that the wire side is only the wrong side when speaking of machine-made papers. In hand-made paper it is the right side. This is rather odd, for undoubtedly the top side would give the most suitable surface.

Australian Paper.—It is stated that from experiments made in Melbourne excellent papers have been made from eleven species of Eucalyptus bark without any addition of rags. This substance, which can be obtained in immense quantities, bleaches easily.

Process for Rendering Paper or Cloth Water-proof, and at the same time protecting it from change. Employ an alcoholic solution of the agreeable oil used to perfume Russia leather, and which is obtained by distilling white birch bark. The oil dissolves readily in alcohol, but is no longer soluble after it has once dried and become oxidized to a resin. The thin film of resin formed by impregnating the fabric does not detract from its pliability in the least, and its aromatic odour protects it from insects.

Another Fire and Water-proof Paper.—The manufacture is accomplished by mixing 25 parts of asbestos fibre with from 25 to 30 parts of aluminum sulphate, and the mixture is moistened with chloride of zinc and thoroughly washed in water. It is then treated with a solution of one part of resin soap in 8 to 10 parts of a solution of pure aluminum sulphate, after which it is manufactured into paper like ordinary pulp.

Another Process for Making Water-proof Packing Paper.—The paper is immersed in a solution consisting of $1\frac{1}{2}$ pounds of white soap dissolved in $1\frac{3}{4}$ pints of water, and also $\frac{1}{4}$ pound of gum arabic and $\frac{1}{4}$ pound of glue dissolved in another $1\frac{1}{4}$ pints of water. Mix these two solutions in a warm state. After the emersion the superfluous fluid is expressed and allowed to dry in a moderately warm temperature.

Another Process.—Paper or pasteboard may be rendered waterproof as follows:—Mix four parts of slaked lime with three parts of skimmed milk, and add a little alum ; then give the material two successive coatings of the mixture with a brush and then let it dry.

A Strong Flexible Paper.—One that is impervious to dampness, is produced by taking a paper composed of strong fibres such as manilla, jute, linen, or the like, and of a quality capable of sustaining a tensile strain of no less than 200 pounds per inch, in the direction of its length when made 12 square feet to the pound. When in the process of its manufacture, or after it has been made, it is rendered impervious to water by the application of suitable size. The paper so prepared is then passed through breaking stamps or rollers, so as to render it limp or flexible ; and this may be done either while the paper is yet in the paper-machine, or in a separate machine adapted for the purpose. It sometimes becomes necessary to pass the paper several times through the breaking rolls and sometimes in contrary directions. When the uses to which it is desired to apply the product demand a very smooth surface, the paper thus rendered flexible is passed through calender rolls in order to smooth it.

Another Recipe for Transfer Paper.—Starch, six ounces ; gum arabic, two ounces ; alum, one ounce. Make a strong solution of each separately in hot water ; mix and apply it while still warm to the side of the leaves of paper with a brush. When dry, a second and third coat may be applied in the same manner, the paper having been well-pressed to make it smooth.

Ancient Paper.—From a microscopical examination of the paper from El-Faijune, preserved in the Austrian Museum at Vienna, in the collection known as " Papyrus Erzherzog Rainer," Dr. Julius Wresner has conclusively proved that linen rags were used in the manufacture of paper as early as the eighth and ninth centuries. The fibre is chiefly linen, but there are also traces of cotton, hemp, and animal fibres present. The manufacture of paper out of rags is therefore an Eastern and not a German or an Italian invention, as has hitherto been supposed. Out of five hundred Oriental and Eastern specimens, not a single one was a raw cotton paper. All those that were examined had likewise been " clayed " like modern paper.

Another Remedy for Grease Spots on Paper.— Grease may be removed from paper in the following manner : Warm gradually the parts containing the grease, and extract as much as possible of it by applying blotting paper. Apply to the warm paper, with a soft clean brush, some clear essential oil of turpentine that has been boiled, and then complete the operation by rubbing over a little rectified spirits of wine.

Protection of Paper against Rats.—Waste paper makes a good thing to banish rats from buildings. Soak the paper in water in which oxalic acid in liberal quantities has been dissolved, and then, while wet, ram it tightly into the chinks through which the rats travel. The rats will never come near the place again.

Semi-transparent Paper Bags as used in the candy business are made by lining the interior surface of the bags with a thin film of fine paraffin wax. This is a mineral wax, as wholesome as bees'-wax, and not only never becomes rancid or changed in quality, but the bags treated with it are well adapted for wrapping every kind of perishable produce. They will preserve coffee, meal, sugar, and any other preparation from the atmosphere and from vermin. The use of these waxed bags is rapidly extending through every department of commerce and industry.

Non-poisonous Vegetable Fly-paper.—Powdered black pepper is mixed with syrup to a thick paste, which is spread by means of a broad brush upon coarse blotting-paper. Common brown syrup will answer, but syrup made from sugar is preferable, as it dries much quicker. For use a piece of this paper is laid upon a plate and dampened with water. The paper may also be made directly at the mill by adding sugar to the pulp and afterwards a quarter to one-third of powdered black pepper, and rapidly working it into a porous absorbent paper.

Smooth-Surfaced Paper.—The remarkable finish of American paper is stated to be due to the addition of a mineral substance called "Agalith," which is a silicate of magnesia, something like asbestos in nature and texture, and which is found only in the United States.

Iridescent Paper.—Boil in water eight parts (by weight) of nut-galls, five parts of sulphate of iron, four each of sal ammoniac and sulphate of indigo, and one-eighth part of gum arabic. Wash the paper in this decoction, and then expose it to the fumes of ammonia until the desired result is reached.

The Life of Printing Papers.—A German scientist

has been making some careful examinations into the composition and durability of printing papers. He appears to have been requested to go into the investigation by the Prussian ministry, who have become concerned in regard to the durability of important state documents. The question put was whether the printed matter of to-day will be in existence fifty years hence. The question would appear to be almost foolish on the face of it, and yet the proportion of printing papers that have fifty years' wear in them is very small. Out of the papers upon which ninety-seven periodicals were printed, only three were perfectly satisfactory; while it was estimated that the paper of thirty-one others contained so much cellulose of wood and straw, and so much mineral matter, that they were hardly likely to last more than half a century. Of the remainder of the periodicals, the paper was still less satisfactory.

Enamel Surface for Paper or Cards.—The following

formula for the brilliant white satin enamel, applied sometimes to French cards and *papier de luxe*, is given in a foreign journal:—For white, and for all pale and delicate shades, take 24 parts by weight of paraffin, add thereto 100 parts of pure kaolin (China clay). very dry, and reduce to a fine powder. Before mixing with the kaolin, the paraffin must be heated to fusing point. Let the mixture cool, and it will form a homogeneous mass, which is to be reduced to powder, and worked into a paste in a paint-mill with warm water. This is the enamel ready for application. It can be tinted according to fancy.

Blotted Notepaper is simply a sheet of note smothered

with blots of all shapes and sizes. The paper is of various tints, and if say a light blue, then the blots are printed in a blue one shade darker than the paper. So the changes are rung, red on red, green on green, and so on. If black on white, then the blots are stippled a very light grey, or worked solid in a black ink greyed down with flake white, which gives the appearance of blotting paper having been used.

Splitting a Sheet of Paper.—It is said that this can be done by laying the sheet of paper on a piece of glass, soaking it thoroughly with water, and then passing it smoothly all over the glass. With a little care the upper half of the sheet can be peeled off, leaving the under half on the glass. Let this dry and it will come off the glass very easily. Of course the glass must be perfectly clean.

A Simple Test for Printing Paper.—Apply the tongue for sizing, and compare opposite sides together for equality of surface. Look through against strong light for spots, and note whether the paper be "regular." Printing paper ought to "rattle" well, and have good strength and surface. When there is a great "rattle," and if the paper has a glistening brilliancy of texture, then most likely straw is present in the fibre, which, when introduced in excessive quantities, causes the paper to break when folded. The paper should, therefore, be creased and then examined.

Paper from Moss.—The U.S. consul at Christiania has made a report in reference to the manufacture of paper from a white moss growing abundantly in Norway and Sweden. Only the mouldering remains of the plant are used, the living growth never being interfered with, and the paper made from this decayed vegetation is represented as being of unusual strength and superior to paper made from wood. If this discovery is all that is claimed for it, it cannot fail to lead to important results, because paper made entirely from wood-pulp is not strong enough without an admixture of cotton waste or rags to give it fibre. Moss has more fibre in it than wood, and is, therefore, not in need of fibrous reinforcement to impart to it the necessary consistency. Not being used until dead, the moss in question is by nature deprived of the soft and aqueous portions, and only the tough veins or fibres are gathered for the paper mill. As the tested moss may be had for the gathering in Norway and Sweden, it is cheaper than rags. Any cheap fibre that will add to the strength of wood-pulp is sure of careful examination from papermakers.

The Processes for making Wood Pulp.—These are now fewer than half-a-dozen: The Mitscherlich, using bi-sulphite of lime; the Francke, in use at Mölndal, near Gothenborg (Sweden), bi-sulphite of lime, in rotary cylinders of steel lined with lead; the Eckmann process, used at Bergvik (Sweden), bi-sulphite of magnesia, in vertical cylinders; the Graham process, used in England, uses sulphite of lime and sulphite of magnesia, with a simultaneous treatment of bi-carbonate of lime and magnesia, and then treatment with sulphureous acid in close vessels. In addition to these we have the latest process of Zahony and Kellner, used at Goritz, in Austria, and which is said to give excellent results. Bi-sulphite of lime is used, but there are great advantages claimed in the complete utilization of the sulphureous acid in the doing away of steam or sulphuric acid in the bleaching. The method has already been adopted in several Austrian mills.

An Early Experiment in Wood Pulp as a Paper comes to light in the following from the "New York Magazine" of September 16, 1795: "A very interesting discovery has lately been made in the State of Pennsylvania, in the art of paper-making, by a Mr. Biddis. It is likely to reduce the price of that important article by producing a saving of rags. The invention consists in reducing sawdust to a pulp, mixing it with the pulp of rags, and forming the paper from this mixture. We have seen a specimen of paper made in this manner, certified to be composed of one-fourth of sawdust, the remainder of rags. The body and surface of the paper appear as good as usual; colour verges a trifle towards a greenish yellow, which we think could be effectively remedied by indigo. We understand that in a paper of a coarser kind a great proportion of sawdust may be used, even in some as far as three-fourths. Mr. Biddis has erected a mill upon the principle of his invention, and taken out a patent, a right to which he proposes selling to one person in each of the states. The sawdust of all our woods may be used for the manufacture, though some are preferable to others."

Æsthetics and Blotting-paper.—A blotting book craze has been raging in New York. The idea is to decorate the walls of a room with them. An æsthetic writer gives directions for making them, thus : " There are several colours of blotting-paper from which to select the shade desired. but cream or buff is the prettiest. Place together four or six leaves for one book. Before tying them together, however, mark, half an inch from the top, two places for slits, which must be cut with a sharp knife. The distance between the slits should be about two inches, and each slit should be at an equal distance from each side of the leaf. After one has been accurately marked and cut, the others must be done the same way. Run through them a ribbon, the colour of which should harmonize well with the design to be painted on the cover. and tie in a bow with ends. The ribbons must be run through the slits from the back. The cover should then be decorated with a design of any kind which may be pleasing. One, for instance, is a broken pen, from which several little pigs are making their escape with great rapidity."

Blotting Paper for Duplicating Copies.—To a German chemist is due an excellent wrinkle by which the old " graph " processes are superseded in a cleanly and efficient manner. Four parts of glue are soaked in five of water and in three of ammonia liquor until the glue is softened. The glue is then warmed until dissolved and three parts of granulated sugar and eight of glycerine are added, the whole being stirred and brought to boiling point. While hot the mixture is painted with a broad brush on thick blotting paper until thoroughly soaked and a thin coating remains on the surface. After a day or two's drying it is ready for use. The writing to be copied is written with aniline ink on ordinary writing paper. Before transferring, the blotting paper is damped with a sponge or brush and allowed to stand for two or three minutes. The writing paper—writing downwards—is placed on the blotting and the air bubbles pressed out. After half-a-minute's pressure the writing paper is removed. The copies are made in the ordinary " graph " manner from the blotting paper and when the impressions grow faint the surface of the blotting is again damped.

New Cutting Machine.—A Chicago firm has patented
a new paper-cutting machine so arranged as to do away with
"cutting sticks." This machine has a flat-edge lower blade
placed in the table bed. The upper knife on descending
makes a shear cut; the knife passes below the table bed and
throws part of the front table out as the knife ascends; the
front table is thrown back automatically, thus making a solid
surface of the table. The set-screws generally used to take
up the wear in the knife bar are dispensed with, and a simple
bevel gib is screwed up by one nut on top, always keeping the
knife bar firm and true. There is no dulling of the knife, as
the knife-dulling surfaces are dispensed with; it saves the
expense of the knives, saves grinding, sticks, etc., which are
sources of annoyance and expense on other cutters.

Wrapping Paper for Metals.—Paper specially pre-
pared for wrapping metallic articles liable to tarnish is made by
sifting on the sheet while in process of manufacture, and before
pressing or drying, a metallic zinc powder known in commerce as blue powder, to the extent of one half the weight of
the dried paper. The sheet is then run between the press
rolls and over the drying cylinders in the ordinary manner.
The zinc powder will adhere to the paper and be partly incorporated with it in greater or less quantity as the sheet of
paper pulp is more or less thick or more or less wet. The
paper may also be sized and then dusted with the zinc powder,
or the zinc powder may be mixed with size, starch, etc., and
then be applied to the surface of the paper.

A Sheet of Paper.—The mere placing of dry fibres one
on another, however long and strong and however intimately
mixed, will not make a sheet of paper. A sheet of strong
paper can only be produced by fibres possessing the quality of
losing their elasticity in water as they become soft and recover-
ing it again when dry. When moist the fibres settle down in
every conceivable direction, forming a confused interlaced
mixture, and in drying each one recovers its original elasticity.
The longer the fibre and the more intricate the mixture the
stronger will be the sheet of paper.

N

Printing-papers with Water-marks.—Makers of good and high-class writings, both machine and hand-made, have been in the habit ever since paper was manufactured of water-marking them with a device, initials, or name in full.   The result is that while in their own interests makers have been compelled to keep up the qualities, the consumer feels sure of getting a first-class article.   In printing papers, on the contrary, it is very seldom indeed that we find a water-mark of any kind, and the consumer hardly knows what he is buying.   There is really no sufficient reason why well-known and reliable makes of printing paper should not be systematically water-marked.   The one objection is that the water-mark would often appear out of the centre, and in some cases perhaps divided in the middle.   This, however, is of no practical importance, as may be judged by an examination of a sheet of "The Times" newspaper, on the edge of which the maker's name, or part of it, always appears.

Manufacture of Paper Pads.—The object of this invention is to bind together sheets of paper forming a pad in such a manner that no binding composition will adhere to the sheets when removed from the pad ; and it consists, essentially, in placing on the edge of the paper to be bound a piece of linen or other porous material and covering the outer surface of the piece of material with a coating of glue or paste, which will soak through and cause the material to adhere to the edge of the paper, so as to bind the sheets together as desired.   This binding is said to be much stronger than the old style. The same effect might be produced by coating one side of the material, and when it is to be used damping or heating it, so as to make it moist and cause the material to adhere to the edge of the paper when pressed against it.

Another Safety Envelope.—The best way in which to secure the inviolability of an envelope is so to cut the flap as that the end of it may turn over on to the right-hand corner of the face of the envelope—where the postage stamp will hold it securely.   We believe that the plan has already been patented in this country.

Origin of Blue Writing Paper.—A singular story has
been told concerning the origin of blue-tinted paper now so
much in vogue for commercial uses. The wife of an English
paper manufacturer named William East, going into the factory on the domestic wash-day with an old-fashioned blue bag
in her hand, accidentally let the bag and its contents fall into a
vat full of pulp. She thought nothing of the incident and said
nothing about it either to her husband or to his workmen.
Great was the astonishment of the latter when the paper
turned out a peculiar blue colour, while the master was vexed
at what he regarded as gross carelessness on the part of some
of the hands. His wife, wise woman! kept her own counsel.
The lot of paper was regarded as unsaleable and was stored
for four years. At length East consigned it to his London correspondent with instructions to sell it for what it would bring.
The unlucky paper was accepted as a happily-designed novelty
and was disposed of in open market at a considerable advance
in price. Judge of Mr. East's surprise when he received
from his agent an order for a large invoice of the despised
blue paper! Here was a pretty dilemma; he was totally ig-
norant of the manner in which the paper had become blue in
colour, and in his perplexity mentioned the matter to his wife
She promptly enlightened her lord; he in turn kept the simple
process secret and was for many years the monopolist of the
blue commercial paper manufacture.

Asbestos Paper.—Mr. Ladewig has devised a new process of manufacturing from asbestos fibre a pulp and a paper
that resist the action of fire and water, absorb no moisture,
and the former of which (the pulp) may be used as stuffing
and for the joints of engines. The process consists in mixing
about 25 per cent. of asbestos fibre with from 25 to 35 per cent.
of powdered sulphate of alumina. This is moistened with an
aqueous solution of chloride of zinc, is washed with water, and
then treated with a solution of one part of resin soap and
eight or ten parts of water mixed with an equal bulk of sulphate of alumina, which should be as pure as possible. The
mixture should have a slightly pulpy consistency. Finally,
there is added 35 per cent. of powdered asbestos and five to

eight per cent. of white barytes. This pulp is treated with water in an ordinary paper machine and worked just like paper pulp. In order to manufacture from it a solid cardboard, proof against fire and water, and capable of serving as a roofing material for light structures, sheets of common cardboard, tarred or otherwise prepared, are covered with the pulp. The application is made in a paper machine, the pulp being allowed to flow over the cardboard. Among other uses the asbestos paper has been recommended for the manufacture of cigarettes.

Discoloration of Paper.—Experiment has shown that the discoloration is due to the action of light upon the paper containing ligneous substances, such as wood, straw, and jute. When the lignine is removed by chemical means, the effect is not produced. The yellowing is said to be due to a phenomenon of oxidation.

The Protection of Paper and Card Stock.—All paper and card stock should be kept wrapped and covered on both sides, as well as at top and bottom, from dust and smoke, and from the discoloration that takes place. Every time you take a sheet off the top cover the pile again, or the top sheet will be spoiled. Some think it unnecessary to cover printed work, but they are in error, for it is as liable as fresh stock to have the edge discoloured ; and if the work is not to be trimmed after printing, this may prove a serious matter. Never handle either printed or unprinted paper or card with dirty or greasy hands. If you do, you are liable to spoil two sheets. Paper stock, if in cases, should be kept carefully boxed and covered, until used. Flat papers—note, letter, folio, half-medium, demy, and others, ruled or unruled—should be carefully shelved and assorted by weights, colours, and sizes, and labelled. The same applies to cards and envelopes.

Letter and Note Papers.—Confusion is often caused by using the wrong term. Letter papers are quarto, and note papers octavo in size. This should be impressed on the memory.

Colouring and Drying Tissue Paper.—Mr. R.

Crompton, of the well-known and eminent firm, has patented an improved process of treating tissue and other paper in continuous rolls after manufacture with various solutions, chemicals, colouring matter or dyes, and then drying the same. The invention is applicable to the manufacture of tracing, waterproof, and cheque paper, and almost any kind that requires chemical treatment after manufacture. The tissue or other paper is passed in a vertical direction between two pressing rolls, covered with felt or other material suitable, supplied with dye ; this serves to more or less impregnate the paper with the colouring matter whilst removing surplus of same. The paper is then passed unsupported through a drying apparatus arranged close to the pressing rolls.

The Practical Testing of Papers.—Different quali-

ties of paper are tested by various means. The strength is measured by its resistance to tearing. In machine papers the strength and stretching power vary according as the force acts lengthwise or across ; in handmade paper there is little difference. In the former the difference is in the proportion of 2·3, according to the direction of the tearing force. The stretching power acts inversely on the strength, that is, is greater across than lengthwise. In order to test the resistance of paper to the most varied mechanical wear, it is crumpled and kneaded between the hands. After such treatment a weak paper will be full of holes, a strong paper will assume a leathery texture. The test also gives a rough idea of the composition of a paper, much dust showing the presence of earth, impurities, while breaking up of paper shows over-bleaching. The thickness of paper is gauged either by measuring the thickness of a certain number of sheets, or by taking that of a single sheet by means of a micrometer, where the paper is placed between two rules, one fixed and the other movable, acting as a pointer showing the thickness of the paper on a dial. Over three per cent. of ash shows the presence of clay, kaolin, heavy spar, gypsum, etc. Microscopical investigation of paper aims at determining the kind and quality of paper. For this a magnifying power of 150 to 300 diameters suffices, when by colouring the paper with a solution of iodine, a yellow coloration shows the presence of wood fibre, a brown one that of linen, cotton or flax, and no coloration that of cellulose. The determination of the kind and quality of size may be made by boiling in distilled water and adding a concentrated solution of tannic acid—flocculent precipitate shows the presence of animal size ; and by heating in absolute alcohol and adding distilled water, when a precipitate shows the presence of a vegetable size.

The Largest Size of Paper made by Hand.— This is called " antiquarian," and should measure $52\frac{1}{2} \times 30\frac{1}{2}$ inches. There is obviously no special limit to the size of machine-made paper. Of course the breadth is fixed by that of the machine, but the length is practically unlimited while the machine goes and the " stuff" holds out.

To Cut Paper into Three or Five.—Every printer has often been vexed in trying to cut paper into three or five equal parts. If he will simply roll the paper into a scroll until the ends meet twice, then mark the junction point with finger-nail or pencil, the sheet will be divided into three parts. If one-fifth is desired, roll the paper four times.

Japanese Handmade Paper.—The Japanese beat the world for handmade paper. It is especially good for etching, and is greatly in vogue among artists. It is exceedingly durable and highly finished, and prints upon it are very much finer than on any other quality of paper. The handmade parchment paper is made from the inner bark of the sycamore-tree, which has a very tough fibre, and it is beautifully finished. The Japanese Government has in its possession official documents printed on this parchment, which are as good as new to-day, after 1,500 years' wear and tear. The Japanese themselves use this paper for houses, coats, umbrellas, screens, and every imaginable purpose. The manufacture of handmade paper is now being principally conducted by the Japanese Government as a national enterprise.

Granulated Paper.—Take some sheets of strong, unglazed paper; make a mixture of clear starch; strain it through a sieve, and, by means of a brush, spread a layer evenly on the surface of the paper; then leave it in the air to dry. Afterwards place the paper thus treated between wet sheets, in the same manner as if it were india transfer paper, in order that it may become slightly damp. Put a stone on the press, and place the paper on the stone, face upwards; take a cloth, the texture of which is more or less close, according to the grain to be obtained; give a moderate pull at the press, and, finally, leave the paper to dry again in the air. This process, which is quite elementary, places the litho printer in possession of granulated paper ready for the draughtsman.

Paper and Eyes.—A well-known French littérateur writes on paper of a greenish hue in order to save his eyes from the glare of white paper. Yellow writing paper is suggested by some experts, but according to someone else, the best paper for " copy " is whitey-brown.

Chinese Rice Paper.—The bloom and softness of rice paper—which, by the way, is not made of rice, but of pith— have always attracted admiration, but unfortunately this paper is too brittle for decorative purposes. The fault is remedied by laying the paper in a slightly warm weak alcoholic solution of lime, which is drawn off and the paper dried on glass plates, the result being pliability without loss of toughness. It retains its velvety surface.

Drawing Paper, as well understood, must be handmade, for no machine has yet been able to " shut together " the fibres of a sheet of paper with anything like the perfection attainable by the human hand, nor to produce a sheet of equal toughness, or so little liable to warp when damped. The evil attending the use of chlorine and other chemicals in paper, has, says a " Times " correspondent, been of late years so completely overlooked, that Sir J. C. Robinson appears justified in speaking of it as an " unconsidered danger to water-colour art."

A Method of Waxing Paper.—A box provided with steam pipes has an upper depressed and corrugated surface, with gutters fed by a funnel, in connection with a perforated plate, felt sheets, and a roller; the felt is first thoroughly saturated with melted wax paraffin, the temperature raised, and the paper waxed by placing the sheets simply on the felt bed and passing the roller over.

MUCILAGES.

TO Test Bookbinders' Glue.—Glue that will stand damp atmosphere is a desideratum among bookbinders. Few know how to judge of quality except by the price they pay. Price is no criterion; neither is colour. The adhesive and lasting qualities of glue depend upon the raw material from which it is made; for if that is inferior and not well cleansed, the glue will have to be unduly charged with alum or other antiseptic to make it keep during drying. Weathered glue is that which has experienced unfavourable weather while drying. To resist damp it should contain little saline matter. When buying the article, venture to apply the tongue; if it tastes salt or acid, reject it for any but the commonest purpose. This tasting will also bring out any bad smell the glue may possess. Another good test is to soak a weighed portion of dry glue in cold water for twenty-four hours, then dry again and weigh. The nearer it approaches its original weight the better it is.

Cement Proof against Boiling Acids.—May be made by a composition of india-rubber, tallow, lime, and red lead. The india-rubber must first be melted by a gentle heat, and then 6 to 8 per cent. by weight of tallow is added to the mixture while it is kept well stirred; next, dry slaked lime is applied until the fluid mass assumes a consistence similar to that of soft paste; lastly, 20 per cent of red lead is added in order to make it harden and dry.

Glue, Paste, or Mucilage.—The following formula is published for making a liquid paste or glue from starch and acid. Place five pounds of potato starch in six pounds of

water, and add a quarter of a pound of pure nitric acid. Keep
it in a warm place, stirring frequently for forty-eight hours.
Then boil the mixture until it forms a thick and translucent
substance.   Dilute with water, if necessary, and filter through
a thick cloth.   Another paste is made from sugar and gum
arabic.   Dissolve five pounds of gum arabic and one pound of
sugar in five pounds of water, add one ounce of nitric acid,
and heat to boiling ; then mix the above with the starch paste.
The resultant paste is liquid, does not mould, and dries on
paper with a gloss.   It is useful for labels, wrappers, and fine
bookbinders' use. Dry pocket glue is made from twelve parts
of glue and five parts of sugar.   The glue is boiled until entirely dissolved, the sugar dissolved in the hot glue, and the
mass evaporated until it hardens on cooling.   The hard substance dissolves rapidly in warm water, and is an excellent
glue for use on paper.

To Make an Elastic Mucilage.—To 20 parts of
alcohol add 1 part of salicylic acid, 3 parts of soft soap, and 3
parts of glycerine.   Shake well, and then add a mucilage
made of 93 parts of gum arabic and 180 parts of water. This
is said to keep well and to be thoroughly elastic.

Transparent Cement.—Mix in a well-stoppered bottle
10 drachms of chloroform with $12\frac{1}{2}$ drachms of non-vulcanized
caoutchouc in small pieces.   The solution is easily effected ;
when finished, add $2\frac{1}{2}$ drachms of mastic and let the whole
macerate from eight to ten days, but without heat.   A perfectly white and very adhesive cement is thus produced.

Insoluble Glue.—The addition of 2 per cent. of potassium
bichromate to the water in which glue is dissolved, just prior
to its use, and exposing the glued article to light, will make it
insoluble in hot water.

Waterproof Shellac or Varnish that will stick the
edges of paper together may be made with a quarter of an
ounce of crude gutta percha dissolved in carbon disulphide to
the consistency of mucilage.

Paste for Mounting Drawings or Photographs.
—A mixture of gum tragacanth and gum arabic forms with
water a thinner mucilage than either of these two gums alone.
Rice flour is said to make an excellent paste for fine paper
work. A solution of two ounces and a half of gum arabic in
two quarts of warm water is thickened to a paste with wheat
flour ; to this is added a solution of alum and sugar of lead,
one ounce and a half each, in water. The mixture is heated
and stirred until about to boil, and then cooled. It may be
thinned with a gum solution.

To Prevent Glue turning Sour.—Salicylic acid will
prevent bookbinders' glue from turning sour.

Dextrine as a Substitute for Gum on Envelopes.
—It can be procured of any chemist or drysalter, and is prepared for use by mixing with boiling water until it assumes
the required consistency. It should only be made in quantities
sufficient for immediate use, as it is somewhat difficult to
re-melt.

Liquefiable Sealing-Wax.—Heat two parts of Venetian
turpentine and dissolve therein four parts white shellac ; remove the heat, allow to cool somewhat, and add 10 parts 96
per cent. alcohol. Rub five parts cinnabar into a paste with
alcohol and add this to the mixture, stirring constantly during
the addition. The whole is put into convenient bottles, and
whenever it is desired to use the wax the preparation can be
made perfectly fluid by immersing the bottles in warm water
and shaking.

Cement for Rubber and Metal.—Take pulverized
shellac dissolved in ten times its weight of pure ammonia. In
three days the mixture will be of the required consistency.
The ammonia penetrates the rubber, and enables the shellac
to take a firm hold, but as it all evaporates in time the rubber
is immovably fastened to the metal, and neither gas nor water
will remove it.

Recipe for Sticking Writing Paper Pads To-gether.—A quarter of an ounce crude gutta percha ; dissolve in bisulphate of carbon to the consistency of mucilage. Apply with a brush to the edges of the paper where required.

Cement Stopping for Wood.—Convert a quantity of sawdust, of the same kind of wood as that of the work, into pulp, by boiling and lengthened immersion in water. When quite soft and pulpy, strain off the water through a cloth, and squeeze the moisture from it ; keep this for use, and when wanted mix a sufficient quantity of thin glue to make it into a paste ; rub it well into the cracks, or fill the holes with it. When it is quite hard and dry, it may be cleaned off and finished. If the work has been carefully done, the patches will hardly be detected.

Stick-fast Mucilage.—The following is said to be a good, well-keeping preparation :—Gum arabic, 8 ounces ; water, sufficient; sulphate of cinchona, 24 grains ; oil of cloves, 3 drops ; glycerine, 4 drachms ; alcohol, 1 drachm. Dissolve the gum in enough water to form a mucilage of proper thick-ness ; add to it the glycerine, and finally the oil of cloves and cinchona sulphate dissolved in the alcohol.

Dry Pocket Glue is made of twelve parts of good glue and five parts of sugar. The glue is boiled until it is entirely dissolved ; the sugar is then put into the glue, and the mass is evaporated until it hardens on cooling. Lukewarm water melts it very readily, and it is excellent for use in causing paper to adhere firmly, cleanly, and without producing any disagreeable odour.

Paste for Labels.—For adhesive labels dissolve one and a half ounces of common glue, which has lain a day in cold water, with two ounces of candy sugar and three quarters of an ounce of gum arabic, in six ounces of hot water, stirring constantly till the whole is homogeneous. If this paste be applied to labels with a brush and allowed to dry, they will then be ready for use by merely moistening with a wet finger.

Marine Glue.—" Liquid marine glue " is a compound prepared by digesting one part of finely-cut caoutchouc, during about ten to fourteen days, with ten parts of oil of turpentine at a gentle heat and under frequent agitation. This liquid is used for rendering wood, ropes, tissues, etc., waterproof, by applying one or more coats. What is usually known as " Marine Glue," without the distinction " liquid " or " solid," is prepared in exactly the same way, and then adding to the solution two parts of shellac—or better, of asphalt—for every one part of caoutchouc employed. The mixture is heated in an iron pot until it has become completely homogeneous, and does not give off volatile vapours. During the heating the mass must be carefully stirred, and the temperature should not be allowed to exceed 140° C. (284° F.) It is then poured out into capsules. For use one of the latter is heated on a water-bath until the contents are melted, when it is transferred to a sand-bath and cautiously heated to nearly 140° C. The edges to be cemented together must be warmed, coated with a thin layer of the marine glue, and then firmly stuck together.

Glue.—A glue ready for use is made by adding to any quantity of glue common whisky, instead of water. Put both together in a bottle, cork it tight, and set it for three or four days, when it will be fit for use without the application of heat. Glue thus prepared will, it is stated, keep for years; at all times fit for use, except in very cold weather, when it should be set in warm water before using. To obviate the difficulty of the stopper getting tight by the glue drying in the mouth of the bottle, use a tin vessel with the cover fitting tight on the outside to prevent the escape of the spirit by evaporation. A strong solution of isinglass made in the same manner is an excellent cement for leather.

Mouth Glue is made by dissolving 1 lb. of fine glue or gelatine in water, and adding $\frac{1}{2}$ lb. of brown sugar, boiling the whole until it is sufficiently thick to become solid on cooling. Pour into moulds, or on a slightly greased slab, and cut into pieces when cool.

Gumming.—The following is the method used in gumming by label-printing firms. The principal thing to be observed is to prevent the sheets from curling while drying; to do this, the gum must be invariably dissolved in *cold* water. To make gum for thick " drapers' " labels, dissolve one pound of gum arabic, at 6*d.* per lb., in one quart of cold water, and strain through flannel. To make gum for chemists' labels, take one pound of gum arabic (at 6*d.* per lb.) and dissolve it in three pints of cold water; add one tablespoonful of glycerine and two ounces of honey. Strain through flannel, and apply with, say, a 5*s.* piece of Turkey sponge, which will last in constant use three or four months: common sponge goes to pieces almost directly. Lay the sheet to be gummed on a flat board, and gum over evenly. Then lay the sheet, gummed side up, on another thin flat board, and place it in the board-rack to dry, and the sheet will not curl. The board-racks used in London generally hold seventy boards each; they are simply made, and not expensive, being similar in construction to a printer's case-rack; but the fillets to hold the boards (which are very thin) are only one quarter the thickness, and of course are placed together as closely as possible, that is to say, as is consistent with the boards working freely in and out.

Mucilage for Pasteboard.—Melt together equal parts of starch and gutta percha. To nine parts of this add three parts of boiled oil, and one-fifth part of litharge. Continue the heat and stir until a thorough union of the ingredients is effected. Apply the mixture hot or somewhat cooled, and thinned with a small quantity of benzole or turpentine oil.

Wafers.—Mix fine flour with the whites of eggs, isinglass, and a little yeast. Mingle the materials and beat them well together. Spread the latter, making thin with gum water, on even tin plates, and, after drying them on the stove, cut to required shape. The colours are imparted by using for *Red*, a little Brazil or vermilion; *Blue*, indigo or verditer; *Yellow*, turmeric, gamboge, or saffron.

To Prevent Gum turning Sour.

—Add a few drops of oil of cloves, or of alcohol, or any essential oil. Five or six drops to a quart of gum is sufficient.

To Keep Mucilage.

—The best way to preserve mucilage is to cork the bottle with an india-rubber stopper. The india-rubber will not stick to the glass as an ordinary cork does : it fits tightly, so preventing access of air from decomposing the gum, and can readily be cleansed. Once used, an india-rubber cork will always be used in the mucilage bottle.

To Prevent Dampness and Brittleness in Making Pasteboards.

—Use gum shellac, 3 parts ; caoutchouc (india-rubber), 1 part, by weight. Dissolve the rubber and shellac in separate vessels in ether free from alcohol, applying a gentle heat. When thoroughly dissolved mix the two solutions, and keep in a bottle tightly stoppered.

Liquid Glue.

—It is not only acetic acid which may be used for preparing liquid glue, but also certain other acids, the most usual being nitric. Proceed as follows :—Break up one pound of good common glue into small pieces and pour upon it one pint of water, taking care that the whole of the glue shall in turn be brought in contact with the water, so that it may become uniformly soft. Then melt in a covered vessel on a water-bath, cool it, and add 3 fl. oz. of nitric acid of spec. gr. 1.335, in small portions, stirring well after each addition. Finally, put it in bottles. This glue will not gelatinize, and will retain its full adhesive power. It is, however, very acid, and cannot be used where acids would be injurious. If the nitric acid were to be neutralized, the glue solution would lose most of its adhesiveness. If oxalic acid is used in place of nitric, and a gentle heat be employed while the acid acts on the glue, the result is the same as with nitric, so far as adhesiveness is concerned. But there is this difference, that the oxalic acid may be removed by lime, and the residuary liquid glue will be found to have lost none of its adhesive property.

Cement for Leather or Cloth.—One pound of gutta percha, four ounces of india-rubber, two ounces of pitch, one ounce of shellac, and two ounces of oil. Melt altogether and use hot.

Diamond Cement.—Soak isinglass in water till it is soft ; then dissolve it in the smallest possible quantity of proof spirit, by the aid of a gentle heat ; in 2 oz. of this mixture dissolve ten grains of ammoniacum, and whilst still liquid add half a drachm of mastic dissolved in three drachms of rectified spirit ; stir well together and put into small bottles. When required for use it must be liquefied by plunging into hot water and applied directly.

Durable Paste.—To make paste that will keep a year, dissolve a teaspoonful of alum in a quart of warm water. When cold, stir in as much flour as will make it the consistency of thick cream, being particular to beat up all the lumps, stir in as much powdered resin as will stand on a sixpence, and pour in a few drops of oil of cloves to give it a pleasant odour. Have on the fire a teacupful of boiling water ; pour the flour mixture into it, stirring well all the time. In a few minutes it will be like mush. Pour into an earthen dish ; let it cool ; lay a cover on and put it in a cool place. When needed for use take a small portion and soften with warm water.

A Recipe for Gumming.—Dissolve a pound of good gum arabic in three pints of cold water. Then add a tablespoonful of glycerine and two ounces of honey. Strain the mixture through flannel. The glycerine prevents the gummed labels cracking and curling up when dry. A sponge is the right thing to use—not a brush. If the mixture is to stand any time, a few drops of sulphuric acid will prevent its turning mouldy or losing strength.

Mounting Labels on Cans.—Paper labels may be made to stick to tin by first brushing the tin over with hydrochloric acid.

To Preserve Gelatine.—Carbolic acid will keep gelatine from decomposing.

Fluid Glue.—A good one ready at all times for instant use, is a most useful article of stock. To make such a glue, melt three pounds of glue in a quart of water, and then drop in gradually a small quantity of nitric acid. When this ingredient is added, the mixture is to be taken from the fire and allowed to cool. Glue so prepared has been kept in an open bottle for two years.

Another Liquid Glue.—Take some good strong glue and mix it with full proof whisky. Let it digest for three or four days, and it will be ready for use.

Gum for Backing Labels.—Take any quantity of clear pure dextrine and mix it with boiling water until it assumes the consistency of ordinary mucilage. Apply thinly, with a full-bodied, evenly-made, and wide camel's hair brush. The paper should not be too thin or unsized. The preparation will dry quickly, and adhere when slightly wet. No more of the dextrine should be mixed at one time than can be used at once, as it cannot be remelted easily.

Paste without Boiling.—An American patent has been granted for an adhesive paste, consisting of flour with an alkali. If the flour be mixed with an alkali in specified proportions in the form of powder a compound is formed which, when mixed with water, will soon assume the consistence of a paste and will become soluble in water. The action of the alkali on the flour bursts the starch cells and digests or dissolves it, increasing its bulk and reducing it to a paste, which may be thinned by the addition of water or thickened by the addition of more of the alkali and flour. This compound will be sold as a powder to be mixed with water by the user. Paste compounded in this way is apt to discolour paper, especially when the paper is coloured or tinted. It is proposed, therefore, to use as a neutralizing agent ammonium sulphate $[(NH_4)_2SO_4]$ or other similar unstable compound, which, when exposed to the alkaline solution, will cause a double decomposition, whereby

o

potassium sulphate or sodium sulphate, as the case may be, and ammonium hydrate are produced. The alkali quickly digests the flour, in a few minutes swelling it up and reducing it into a paste, which gradually grows thicker in consistence. At the same time, but more slowly, the ammonium sulphate acts upon the strong alkali, and a double decomposition takes place, which results in neutral potassium sulphate or sodium sulphate, as the case may be, and ammonium hydrate, which latter passes off slowly as a gas, thereby gradually decreasing the alkalinity of the mixture. This reaction is slow, and therefore the ammonium sulphate does not neutralize the strong alkali until after the flour has been completely digested. After the completion of this process, the paste, if not of the desired consistence, may be made more fluid or thicker by the addition of water or of the paste compound. After the paste is mixed it is free from lumps, and it will remain in condition for use for several weeks without spoiling. As a powder, it may be preserved without change indefinitely if kept in a dry place, and the fact that a given quantity of the powder will make about eight times its own volume of paste makes it easy to handle commercially and to transport.

Another Method for Pasting Labels on Metal.—It is well known that paper pasted, gummed, or glued on to metal, especially if it has a bright surface, usually comes off on the slightest provocation, leaving the adhesive material on the back of the paper with a surface bright and slippery as ice. To overcome this it is suggested that the metal be first dipped into a strong and hot solution of washing-soda, afterwards scrubbing perfectly dry with a clean rag. Onion juice is then to be applied to the surface of the metal, and the label pasted and fixed in the ordinary way. It is said to be almost impossible to separate paper and metal thus joined.

To Preserve Paste.—The decomposition of paste may be prevented by adding to it a small quantity of carbolic acid. It will not then become offensive, as it often does when kept for several days, or when successive layers of paper are put on with paste.

Cement for Envelopes.—" Inviolable " envelopes, long the desideratum of envelope makers and users, may be rendered thus by using a solution of cupric ammonia. The latter dissolves cellulose and other substances. If the envelope is moistened with the solution, the surface of the paper is to some extent disintegrated, and a joint is effected which, when dry, can be opened only by using some force.

Mucilage of Acacia, or Gum Arabic.—It is usually stated that the preparation of this mucilage in the cold renders it less liable to become thick and muddy. This the " Pharma. Zeitung," in a late issue, disputes upon the authority of H. Notfke, who recommends the hot treatment. The following is his plan : " By repeated experiments I have become convinced that the keeping qualities of mucilage gum arabic are improved by using hot water for solution. The water is first heated to boiling, then allowed to cool to about 80° C. (176° F), and this is then poured upon the gum arabic, which had previously been carefully washed with cold distilled water. The whole is well covered, set aside in a cool place, and frequently stirred. Solution will take place rapidly. Any water lost by evaporation is replaced before straining. The straining must be done by passing the mucilage through a woollen strainer, previously washed repeatedly with distilled water. The strained mucilage should be filled into small bottles (2 to 8 oz.), which had previously been carefully dried in a drying oven, and must be still hot when filled. They should be filled to about three-fourths of the neck and at once corked, the cork, if possible, being pushed down so as to meet the surface of the mucilage. Mucilage thus prepared, is clear and bright, and, if put up as here directed, keeps well, though it generally becomes faintly opalescent after a few days. But this happens also to mucilage prepared by the cold process. If the mucilage is heated in a steam bath, immediately after it has been strained, it will keep still better, but this second heating renders it quite opaque, which is not considered proper."

To make Wall Paper Stick.—To make paper stick to a wall that has been whitewashed, wash in vinegar or saleratus water.

Flexible Compound for Paper Pads.—Use one part by weight of sugar, one part of linseed oil, four parts of glycerine, eight parts of glue or gelatine, a little aniline dye to give colour. Cover the glue or gelatine with water and soak one half hour or until soft. Pour off all the water and melt by heating in a pail or basin placed in another kettle containing boiling water ; a common glue-pot will do. After melting, put in the sugar and glycerine, remembering to stir well ; add the dose and then stir in the oil thoroughly. Green and carmine are good colours, and when both are used a handsome purple will be obtained.

Moisture-proof Glue.—Dissolve sixteen ounces of glue in three pints of skim milk, and if desired still stronger, add powdered lime. For marine glue, heat moderately a mixture of india rubber (one part by weight), mineral naphtha or tar (two parts), and add twenty parts of lac in powder. Heat to a temperature of 120 degrees to use.

Ordinary Sealing Wax.—The basis of all the different varieties of sealing wax is shellac and Venice turpentine. Fine red sealing wax is made as follows :—Melt cautiously 4 oz. of very pale shellac in a bright copper pan over a clean charcoal fire to the lowest temperature of heat that will be necessary to melt it. When the shellac is melted stir it into $1\frac{1}{4}$ oz. of warmed Venice turpentine, and then 3 oz. of vermilion. The heat should be so maintained as to be just sufficient to permit of the thorough incorporation of the constituents. When this latter has been effected the fluid mass is discharged into metallic moulds and allowed to cool. To produce the beautiful polished surface of the sealing wax of commerce the sticks are removed from the moulds in which they are cast and placed in other moulds of polished steel, which may be engraved with the name or brand of maker, and with any desired ornamentation. The steel moulds are heated just enough to melt the surface of the sealing wax, and the sticks thus acquire, when cooled, a beautiful glossy surface. Different colours are obtained by the incorporation of suitable colours instead of the vermilion as above. To perfume sealing wax add to the ingredients, when somewhat soft, and thoroughly incorporate with them one per cent. of liquid storax or balsam of Peru, but a little musk essence or amber-gris will answer the same purpose.

Substitute for Gum Arabic.—A substitute for gum arabic, which has been patented in Germany, and is likely to be largely used for technical purposes now that good gum arabic is so scarce, is made as follows :—Twenty parts of powdered sugar are boiled with 7 parts of fresh milk, and this is then mixed with 50 parts of a 36 per cent. solution of silicate of sodium, the mixture being then cooled to 122° F. and poured into tin boxes, where granular masses will gradually separate out, which look very much like pieces of gum arabic. This artificial gum copiously and instantly reduces Fehling's solution, so that, if mixed with powdered gum arabic as an adulterant, its presence could be easily detected. The presence of silicate of sodium in the ash would also confirm the presence of adul-teration.

STEREOTYPING AND ELECTRO-
TYPING.

STEREOTYPE Moulds.—To make casts or moulds of plaster of Paris from metal types, without being troubled with air bubbles or picks, as they are sometimes called, use the finest and purest plaster of Paris obtainable. When filling a mould, beat up the requisite quantity of cream quickly and with care, to avoid making it too thick. In pouring this in, use a good camel's hair brush to displace air bubbles; a mere surface cover of this thin cream is all that is requisite. While doing this, have ready the thicker plaster, of the consistency of light syrup, and fill up the mould at once. In about twenty minutes open the mould, if the plaster is pure and has been properly mixed. If too much oil is not put on the type, and the brush is used properly, it will result in clear, sharp moulds.

Composition for Mounting Plates.—The following is recommended for fixing electros on wood, etc. Common joiners' glue is dissolved to a consistency of syrup, and pure wood-ashes are added under constant stirring until the mixture has the appearance of varnish. The adhesive power of this composition is said to be very great, the addition of the ashes preventing the electros from parting from the wood even when washed with lye.

Another Method.—A good mastic for fixing stereotype plates on wood or metal is obtained by dissolving ordinary cobblers' wax until it is about the thickness of a syrup, and then stirring in a sufficient quantity of wood-ash to make it into a kind of varnish.

How to Preserve Electrotypes.—When electrotypes are out of use and require to be stored, they should be kept in a dry place, and the surface of the plates should be oiled in order to prevent verdigris. When they become clogged with hard, dry ink, which the pick-brush and turps fail to remove, they may be cleaned and made equal to new in a few minutes by covering their surface with a little creosote, and afterwards brushing the surface with turps.

Nickelled Electrotypes.—There are now being produced a new species of nickelled electrotypes. The electric current is produced by a Gramme machine ; the mould is of caoutchouc, and after being taken and blackleaded, is put in a bath of nickel, where the electric current coats it with a thin covering of that metal in a short time. The mould is then withdrawn from that bath and placed in one of copper solution, where it remains for some hours. Afterwards the cliché is subjected to the ordinary processes of backing and finishing, and it is then ready for use. These electros cost 10 centimes the square centimètre (about $6\frac{1}{2}d$. the square inch), and are of especial value where vermilion ink is used.

Stereo Metal.—Printers should know that to make sterco metal they should melt together old types and grocers' tea-chest lead, using 14 lbs. of the former to 6 lbs. of the latter. It is preferable to melt the type and the tea-lead separately, and mix them when molten. To prevent any smoke arising from the melting of tea-lead, it is necessary to melt it over an ordinary fireplace, for the purpose of cleansing it, which can be done by throwing in an ordinary piece of tallow about the size of a nut, and stir it briskly with the ladle, when the impurities will rise to the surface and can be skimmed off.

Valuable Hint to Stereotypers.—Occasionally, when it is rendered necessary for engraving purposes to run a type-metal plate in a plaster-of-Paris mould, the metal comes out punctured with numerous air and steam holes, not at all like the smooth surface of the plaster in which it has been cast.

These undesirable results are caused by the molten metal
bubbling like boiling water as soon as it touches the mould,
then cooling and hardening full of cavities on the under side
where it is intended for engraving.   So very objectionable
a result can be readily obviated by a few simple precautions.
Warm the plaster mould, then fasten it face upward in a
shallow cast-iron pan, plunge pan and all into a pot of type-
metal, and keep it immersed until bubbles of air and steam
are no longer given off.   Then draw it out, permit it to cool,
and separate the plate, and it will be found free from cavities
and ready for planing and trimming.

Stereotyping Music.—The principal London music
printers prefer plaster stereos to electros of music formes.
The reason is that the pressure, perhaps as much as fifty
tons, on the wax to get the mould has a tendency to spread
out the types, and to cause bad joinings.   The plaster is
merely poured on, and in drying the mould contracts, which
tends to bring lines together.   Some firms, however, work
from an electro taken from a plaster stereo, and in this way
obtain the advantages of both processes.

Proportion of Antimony for Stereo Metal.—
About 12 per cent. of antimony is sufficient to mix with lead
for stereo metal, and will make it quite hard enough without
the addition of tin.

Stereotypers' Paste.—Take 5 oz. of flour, 7 oz. of
white starch, a large tablespoonful of powdered alum, and
4 quarts of water.   Put the flour, starch, and alum into a
saucepan, and mix with a little of the water, cold, until the
whole becomes of the consistency of thick cream.   Then
gradually add the remainder of the water, stirring well meanwhile to prevent lumps.   Put the mixture over the fire and stir
until it boils; then let it stand until quite cold, when it should
look like jelly.   When you are ready for work add Spanish
whiting, the mixture not to be too stiff to spread readily with
the paste brush.   Put through a fine wire sieve with a stiff
brush, and it is ready for use.

Removing Paraffin from Plates.

—In electro foundries where paraffin is used for cleaning plates, the oil is almost instantaneously removed by immersing the plate in a bag of mahogany sawdust. Nothing so readily acts on the paraffin and brightens the metal. We are afraid, however, that the treatment would not answer for movable type formes.

Celluloid Stereotypes.

—Celluloid is coming into use for stereotypes. The engraving or forme of type to be stereo-typed is first used to make a fine paper matrix, just as if a common metal stereotype was to be made. Then this matrix is placed in a forme, and over it is laid a sheet of celluloid. The two are put in a hydraulic press, the temperature is raised to 300° Fahr., the celluloid is pressed into the matrix at a pressure of 4,000 lbs. to the square inch, and then the thing is done. When taken out and cooled the celluloid plate is an exact counterpart of the original forme, and when cemented to a suitable wood backing it is good for four times as many impressions as a copper stereotype. Besides that, it is not easily damaged. Celluloid is also used for facing wood type by laying a thin sheet of celluloid over the face of a big block of wood, and the two are placed in a hot press. When they come out the celluloid has been forced into the pores of the wood an eighth of an inch, and has a surface as smooth as metal type. The block is then cut up into wood type by the ordinary wood-type machine, or it may be sold to wood en-gravers, who find it equal to boxwood.

Another Method.

—Celluloid stereotypes are made by placing the dry mould and the celluloid of which the stereo-type is to be made, in a frame provided with a spring, which will keep the celluloid under constant pressure. The whole is then immersed in hot oil until the celluloid is sufficiently softened to be forced into the mould by the spring.

Electrotyping Handwriting.

—To produce electro-types or stereotypes of letters, signatures, ordinary written matter, drawings, or sketches, coat a smooth surface of glass or metal with a smooth thin layer of gelatine, and let it dry.

Then write or draw upon it with an ink containing chrome alum ; allow it to dry exposed to light, and immerse the plate in water.   Those parts of the surface which have not been written upon will swell up and form a relief plate, while those parts which have been written upon with the chrome ink have become insoluble in water, after exposure to light.   The relief may be transferred to plaster of Paris, and from this may be made a plate in type-metal.

Alloy which Expands on Cooling.—Most metals and alloys shrink or contract on cooling.   But an alloy which will expand on cooling may be made of lead nine parts, antimony two parts, bismuth one part.   This alloy can be advantageously used to fill small holes and defects in iron castings.

Stereotyping Cold.—A new method is said to have been invented by a stereotyper.   Its sole novelty consists in the mode of drying the damp matrix as taken from the type.   The matrix is laid with its face upon a fine wire gauze, which is fixed in a frame the edges of which adapt themselves to the form of the edges of the matrix.   A second frame is then put upon the first one, and the two fastened together by cramps; the edges of the matrix are consequently fixed firmly between two frames, its face resting upon the wire gauze.   The matrix is then dried by simply placing it in a chamber by the side of the melting-pot, heated by the escaping gases of combustion of the latter.

To Make Flong.—Lay down upon a smooth iron or stone surface a piece of stout brown paper.   Paste the surface of this over equally.   Lay a sheet of good blotting paper upon the pasted surface, and press it down with the hand.   Paste this over, and then put another sheet of blotting paper on ; smooth well, paste this over; then place a sheet of good tissue paper or copying paper upon the blotting paper.   Press it well down again, paste over again, lay another sheet of tissue on the last, and smooth the whole carefully.   Some pass a small steel roller over the flong to incorporate it more together and give it greater firmness. Flong may be had ready prepared; or it may be made with Nicholson's paste in the following manner : Paste a sheet of thick blotting paper evenly with the prepared composition and lay upon it a sheet of tissue. Turn the sheet and paste the blotting paper again, and lay upon it a sheet of thin demy paper. Turn again and paste the tissue side, and lay down a second tissue. Paste again and lay down a third tissue. This flong must not be immersed in *hot* water.

To Mount Stereo Plates.—For fixing stereo plates on type-high stereo cores, thereby saving the expense of the old wood-blocks and also the labour in fixing, the following plan comes from across the Atlantic :—The bed is a plain iron surface of any stereo-block height. It is placed on a steam-chest to be warmed, and is then coated by a brush with cement composed as follows: beeswax, 1 lb. ; gum ibus, 1 lb. ; Burgundy pitch, $\frac{1}{4}$ lb. It is then removed to an iron table to cool. When quite cold, the stereo plates are placed on the dry cement and adjusted. In order to do this accurately, a light wooden frame is laid over the iron bed, with cross threads stretched at proper intervals to mark the margin. The iron bed is then again pushed on the steam-chest, and as soon as the cement is melted, the bed is shunted on to the bed of a press. Loose sheets are laid over the plates to soften the impression, and the table is run under the platen. The pressure is applied till the cement is cold, then the forme is ready for the machine. The plates are got off with a stout knife, or melted off.

Sharp Stereo Plates.—The French obtain very sharp plates in the paper process by using a compound as follows : 1 kilogramme paste made of starch, 1 kilogramme kaolin (china clay), and 10 grammes yellow dextrine. These are well mixed and passed through a hair-sieve or fine muslin. After standing for about two hours the composition is ready for use. The spreading of this mixture is said to be more difficult than that in ordinary use, but the sharpness of the matrices obtained more than compensate for it.

Correct Heat of Stereotype Metal.—Dip a piece

of paper in the metal, and if it turns it black the metal is too hot, but if it turns the paper to a straw colour it is correct.

American Stereotype Metal.—This is compounded

in the United States thus:—100 lb. good lead, 16 lb. antimony, 4 lb. block tin. The quality of stereotype metal may be tested if required without analysis by pouring melted metal on an iron surface. If, in cooling, it becomes a deep bright steel colour, it may safely be inferred that the metal is good. Or if, in cutting the surface with a graver, it seems grainy or gritty as the instrument is passing over it, it may as safely be adjudged sufficiently hard. It is absolutely essential that the metal should be of the proper consistency, because if too hard it is likely to break on the press; if too soft, to mash. It is, however, a difficult, if not an impossible task, to apply an absolute test except by analysis.

Mice-proof Stereo Paper Moulds.—By reason of

the paste used for sticking together the different sheets, it is found that paper matrices or moulds are frequently injured by mice. Risks of this nature can be obviated by mixing with the paste some bitter substance, picric acid being especially recommended for the purpose. Mice will never touch paste that has been thus treated.

Some Defects in Stereotyping to be guarded

against.—We take the following remarks from the " Press News." Plates are bevelled usually by a rule of thumb method, so they are any width within a " lead " or so, and the planing might be almost as well left alone, only the casting-boxes are not true. For instance, the casting-box is cold to start with, and metal is poured into it, with type-high gauges, to make it hot. These " heaters," so called, are put back into the metal-pot, or stood aside, if their going back whence they came would cool the metal too much. Thus the casting-box becomes hot enough to cast a plate in. The " mould " is laid on the box, the proper pica or other gauges laid in their places on its margin, the back of the box is lowered, the cross-piece is swung into its right place, the screw is turned with a spin, and afterwards by the operator's two hands, and the whole is then tilted and the metal poured in. The difference of shape, if any, of the casting-box, owing to its ever-varying temperature, will be hard to alter, but, oh, that screw just referred to! As the box gets hotter its iron gets more pliable ; as the bustling men get hotter they get stronger, and down goes that screw on to the middle of the back of the box, just about as tightly as they can twist it, and the space where the metal is to run in is reduced by this compression, and the plate is thinner there, and the metal cools most there and shrinks as much as it can, so the plate needs planing. If, instead of this mischievous screw, a bar was placed across on the " gib-and-cutter " principle, the mould and gauge would keep their places, the back would not be depressed, and the plate would be truer, With many planes in use to-day, you can take three cuts off a plate without altering any part of the machine, or the " papers " under the plate. This should not be. It is wrong.

To make Glue Stamps for Endorsing.—Someone writes: "I have stopped using rubber stamps and now use glue ones in my business, and for stamping purposes I use letter-press ink. The glue stamp is made in the following manner :— Upon the type or block I place several leaves of tinfoil ; over this a piece of felt is placed and the whole pressed tightly. Loosen the press at once and take out the tinfoil matrix, which is now ready for use. Oil matrix slightly and surround with oiled bridges or reglets ; then pour fluid glue mixed with a little roller composition over it. After it gets cool the layer of the glue will loosen itself slightly from the matrix, and become sufficiently hard but still remain elastic. The glue I use is that used by carpenters. During the first few days the glue may appear somewhat soft, but it soon becomes nice and hard, and retains sufficient elasticity. I am very well satisfied with the stamp, as it prints in an excellent manner."

Moulds and Matrices for Stereotype Plates.—A new invention just patented consists of a dry method in the manufacture of moulds. The patentee, a Mr. Eastwood, of Hull, makes the mould of two parts, viz., a facing and a backing. The face is composed of a piece of muslin covered with a sheet of tissue paper; the muslin is soaked with, and the paper (when more than one sheet is used) is pasted with, a composition which will keep the muslin in a flexible state, prevent the paper drying hard before use, render it sensitive to moisture, prevent contraction on application of heat, and harden the facing when heated. The composition used for this purpose is formed of glycerine and starch, with or without a small quantity of common salt, suitable proportions of which are about 6 ounces starch, 1 ounce glycerine, and 20 grains salt. The backing consists of a dry thick sheet of soft paper—blotting paper, felt, or other suitable substance— capable of receiving and retaining an impression. The facing composed of the muslin and tissue paper is dried cold, and when used should only contain sufficient moisture to render the paper slightly soft. The muslin should be kept firm and wiped with a sponge containing the composition. In taking the matrix the facing is placed upon the type, paper side downwards, and the backing of dry paper is placed upon it. They are then covered with woollen or india-rubber blanketing, and rolled or pressed. The matrix is claimed to be at once formed, and when removed from the type has simply to be warmed through. The heat quickly sets the composition and hardens the face.

Copper-plating on Zinc.—The use of cyanide baths for plating on zinc has the double disadvantage of being poisonous and expensive. A recent discovery has overcome the objections by rendering the cyanide bath unnecessary. This is accomplished by the use of an organic salt of copper, for instance a tartrate. Dissolve 126 grammes sulphate of copper (blue vitriol) in 2 litres of water: also 227 grammes tartrate of potash and 286 grammes crystallized carbonate of soda in two litres of water. On mixing the two solutions a light bluish-green precipitate of tartrate of copper is formed. It is thrown on a linen filter, and afterwards dissolved in half a litre of caustic soda solution of 16° B., when it is ready for use. The coating obtained from this solution is very pliable, smooth, and coherent, with a fine surface, and acquires any desired thickness if left long enough in the bath. Other metals can be also employed for plating in the form of tartrates. Instead of tartrates, phosphates, oxalates, citrates, acetates, and borates of metals can be used, so that it seems possible to entirely dispense with the use of cyanide baths.

New Soldering Mixture.—One has been found which is free from the defects of chloride of zinc. It is made by mixing one pound of lactic acid with one pound of glycerine and eight pounds of water. Chloride of zinc, so much used in soldering, has, besides its corrosive qualities, the drawback of being unwholesome when used for soldering the tins employed to can fruit, vegetables, and other foods.

Patent Stereotyping Millboard.—A German has simplified stereotyping by inventing a sort of stereotyping millboard, which dispenses with preparing and preserving paste, and making the flong. When the millboard is to be used it is put for five minutes in cold water, and then placed between a layer of blotting paper to take off all the water on the outside. It is then laid on the forme and beaten or pressed, where it will dry, giving a sharp impression, within five or six minutes.

Electrotypers' Moulding Wax.—The following proportions are generally used :—
25 lbs. pure commercial beeswax.
1 oz. white turpentine (lump).
$\frac{3}{4}$ lb. plumbago.
The wax must be melted and the others mixed in.

Stereotypers' Improved Paste.—This is composed of the following ingredients : Water, flour, starch, gum arabic, alum, and whiting. The best of flour and starch are to be used. These foregoing articles, excepting the whiting, are thoroughly mixed, and heated by steam. When the mass is thoroughly homogeneous, sufficient whiting is added to give it stiffness.

To Stereotype Zinc Etchings.—A very wet mould, free from creases, is recommended. Those who use ready-made flong should let it lie twenty minutes in cold water, carefully dry between blotting-paper, and then beat in. Those who prepare their own matrices will mix the matrix-powder or chalk with cold water to the consistency of milk, and spread thickly on the paper, beating continually but lightly with a soft brush. All zinc etchings must be thoroughly washed with paraffin before stereotyping, and also woodcuts which are to be stereotyped. Old woodcuts and etchings which are covered with hard, dried ink must be laid face down in paraffin for at least half an hour, otherwise the stereotype will not be a success.

Improved Stereo Core Blocks.—Objections have been made to the various plans of preparing wooden or other cores for stereotype or electrotype blocks. These cores, when not covered by the metal, were liable to warp or swell from the water used in washing the formes. To overcome this fault a new method of making a core has been perfected. The wood. or other material is placed over the casting-box and allowed to remain until sufficiently heated, supports being set or driven into the ends at points previously marked. When a stereotype is to be made, the matrix is inserted in the casting-box as in the old method, and the core is put into position by means of the supports, two of which should be in the lower and one in the upper end of the box. All the pieces of box being in place, the upper support of the core is covered with stereotypers' paste, on the top side only, to prevent its melting when the hot metal is poured against it. In this way a stereotype is made and the core covered in one operation. A different method is pursued when a stereotype or electrotype already cast is to be mounted on a block. First, it is backed in the usual manner, and is straightened or planed. Next, the face is laid against the bottom of the casting-box, the core is put on the back, and strips of tinfoil put around the edges, the intention being for these strips to fuse as the metal is poured. Pieces of fusible metal—one, two, three, or four as required—just thick enough to fill the space between the core and the lid of the casting-box, are put on the back of the core ; then, with the enclosing side and end bars set, the core is ready to receive the metal. If the plates are old or corroded, the edges around the outside of the core, where they are to fuse with the new metal, should be scraped clean and bright, and the strips of tinfoil applied as in new, bright plates. Blocks made in this way have been found, after many practical tests, to be absolutely waterproof and cannot be affected by the air. They also require less metal than the old style partially enclosed cores, with ends and sides exposed to moisture.

LITHOGRAPHY.

HOW Photo-Lithography is done.—When the photographer is required to produce a number of copies of a landscape or a portrait, he first of all makes a "negative," a kind of transparency, in which the parts corresponding to the dark portions of the original are transparent, while those corresponding to the lights are opaque. This negative is placed on a piece of paper so prepared that it goes dark whenever exposed to the light. Accordingly, those parts which are covered by the opaque parts of the negative remain of the original colour, but those on which the light penetrates through are turned to a dark colour. In this way a paper positive is produced—a picture in which the lights and shades are correctly reproduced. The object of photo-lithography is to get a picture transferred to the lithographic stone in fatty ink, so that prints may be struck off in the ordinary lithographic manner. A sheet of plain paper is coated with a solution of which gelatine is one of the principal constituents. The paper is dried, and then soaked in bichromate of potassium, when it becomes sensitive. It is placed under the negative, and when the print has been made, it is entirely covered with printing-ink. It is next soaked well in warm water, and strange though it may appear, when gently brushed over, the ink will be removed from all those parts which were not exposed to the action of light. In fact, there is an image in fatty ink on the paper. To lay it face downward on a litho stone and make a transfer is but the ordinary system of lithography. From the stone may be printed almost any number of " photolithographs."

Lithographers' " Tushe."—Artists and printers using the air-brush find that the dots blown upon the stone are of such a fine texture that they are unable to etch and prepare them safely, and ask for a "tushe" for this process which will stand a stronger etching. A German gives the following recipe, by which artists will be enabled to use the air-brush to still greater advantage :—Take 10 parts of shellac, 10 of bees-wax, 8 of Castile soap, 6 of dragon's blood, 6 of tallow, and 10 of palm oil—total, 50 parts. Wax, soap, tallow, and palm oil are put on the fire in an iron pot large enough to hold at least ten times the quantity of these ingredients. The fire should be slow-burning, but as soon as the ingredients are hot enough light the evaporating gases and burn for a few minutes. Hold a cover ready in one hand, and cover the pot after about two minutes. While burning add the dragon's blood and shellac and stir well. When all ingredients are well melted and mixed let the whole cool off. Form the mass at the right time into balls or sticks, and in using the tushe melt one part of tushe with 10 parts of distilled water in a boiling condition. A tushe, as described, will, if kept air-tight, last in a liquid form for about six months, and will stand a very strong etching.

Metal Back for Lithographic Show-Cards.—A process has been devised by which paper may be caused to adhere to metal with such persistency that no part of it can be removed without destroying it. The primary object of this invention is to enable lithographers to provide metal backs to fancy show-cards or pictures in such a manner that when the picture is once placed upon the surface of the metal it will be-come permanently attached thereto, and not peel off, even when subjected to severe changes in the temperature of the atmos-phere. The process is as follows :—"Any suitable acid would probably answer the purpose : but muriatic acid of full strength, with enough zinc added to it to prevent it effervescing by the addition of any more zinc, and when thus killed mixed with an equal quantity of water, produces the desired result when rubbed over the surface of the metal plate. After thus coating the metal plate, it is placed in an oven to facilitate its drying and when dry it is painted with a coating of fi e varnish—

preferably that kind of varnish used by coach-builders in painting the gearing of their vehicles. This coating of varnish is left to dry until it becomes merely sticky. To facilitate this drying, the plate is put into a japanner's kiln, which, if kept at a temperature of about 110 degs. Fahr., will bake the varnish sufficiently to produce the desired effect in twenty minutes. If not placed in a kiln, it will take about five hours to acquire the desired condition. The sheet of paper at this period should be carefully placed on the surface of this prepared plate, when, by submitting it to heavy pressure in a lithographer's or other suitable press, it will be found that the sheet of paper has become so thoroughly amalgamated with the roughened and prepared surface of the plate that it is utterly impossible to peel it off the plate. With the view of finishing or ena-melling the surface of the paper thus placed on its metal plate, the surface of the paper is; coated with a suitable sizing pre-pared as follows:—One pound of gelatine glue dissolved in water with an ounce of honey makes a suitable elastic sizing for coating the surface of the paper. When this sizing is perfectly dry the surface is painted with a coating of the best dial varnish. The plates thus prepared should be placed in racks and allowed to remain for a couple of days before being used, when it will be found that they can be handled, and the paper with pictures or any ornamentation which may have been printed on it will be incorporated as perfectly with the metal plate as though the painting had been placed directly on the metal."

Printing Dry from Zinc.—Recently the "Lithogra-phische Rundschau" had an important article on this subject. In the introduction the editor says that there are very few branches indeed which have so many secrets, recipes, and miraculous appliances as lithography and its allies. Many of them are offered for sale ; when, however, the money is paid to the "inventor," it is found quite frequently that such re-cipes, etc., were known long ago. This has reference also to secret fluids and tinctures for which it is claimed that when a portion is mixed with the lithographic printing ink we are enabled to print dry from lithographic stone. It is well known that such fluids consist only of glycerine, and the "secret" certainly does enable one to print for some time without damping the stone, but with the result that it spoils the ink and makes the inking rollers slip. To print from stone without damping certainly has great advantages provided it can be practically carried out. In 1885 there was a great deal said in reference to a new kind of lithographic composition roller, for which it was claimed that, by the use of a specially prepared kind of lithographic printing ink, one could print dry from stone. After a while, however, it was found that these composition rollers got out of order and out of shape ; they became uneven, the inking could no longer be done in a solid manner, and it was not long before the rollers were out of order altogether. It was strange that this method of printing dry without damping was of more practical use in printing from a zinc plate than from stone. The uncertainty of the process kept us for a while from further experimenting, but we did not lose sight of it. Recently we had to print autographs which, being in editions of 200 copies only, would not have been profitable had we printed them on a steam-press. Hence, we printed them from zinc on the hand-press without damping the zinc plate, and we succeeded wonderfully. This encouraged us to apply the same to better work, and our own practical experience convinced us that this method of printing dry from a zinc plate without damping is of a value which should by no means be under-estimated. In order to give a practical proof of his experience in this matter the editor inserts in his paper a page printed (without damping) from zinc on the hand-press. It is, indeed, a well-printed page in blue-black ink. In explanation he says, "We added to the blue-black ink some glycerine and a trifle of lard. Every practical printer will soon ascertain the correct proportion. We have no doubt that the same method of printing (without damping) from zinc will work also on a zincographic press."

The Origin of Lithography.—The art of lithography was finally perfected through a series of suggestions made by accident. A poor musician, who could not afford the expense of having his compositions engraved on copper, became curious to know whether the notes and lines could not be etched on a smooth stone as well as on copper. Having smoothed and prepared a stone, his mother desired him to make a memorandum of such clothes as she proposed to send away to a laundress. Not finding a pen, ink, and paper conveniently near, he hurriedly wrote the list on the stone in the etching preparation, intending to make a copy of it later in the day; but being called away, the writing was allowed to remain, and, a week later, when he resumed his experiments, and was about to clean the stone, he wondered what effect a coating of aquafortis would have. He applied the acid with a camel-hair brush, and in a few minutes was amazed to see his writing standing out in relief. All that was then necessary was to moisten the stone with a damp cloth, ink the writing, and take off an impression, and thus the art of lithography was born.

New Preparation for Stones.—A new compound, called "The Universal Preparation," has been introduced in America. It consists of pure gum arabic and a chrome salt, and a few chemicals to keep the preparation in good order. The properties claimed are the following :—That this preparation is not an acid, but so composed that when applied it is impossible to spoil the stone even with the grossest carelessness ; that it will prevent tinting, filling up, or smearing up of the work ; that no double or repeated preparing is necessary, as the work after one application will remain unchanged on the stone, and that by the use of this preparation stoppage of the press to etch or clean the same will be prevented, and the work can be done mechanically, while running at fastest speed, by any pressman or printer.

Prevention of Newly-printed Sheets Sticking together in Chromo-Lithography.—Use as much thin varnish in the ink as may be needed to obtain the lightest possible tint, without smearing the work on the stone, if you add a trifle of silicate basic potassa to the ink ; but care must be taken that it is done drop by drop, as it has a very strong effect, and will make the ink as stiff and strong as if it had been mixed with the strongest varnish.

The Sizes and Weights of Lithographic Stones.

Printing Surface.	Size of Stone in inches.	Average weight in lbs.
Demy 8vo	9 × 7	13
Imperial 8vo	12 × 10	24
Demy 4to	14 × 10	27
Royal 4to	14 × 11	32
Crown Folio	16 × 12	43
Large Post Folio	17 × 11	42
Imperial 4to	17 × 13	49
Demy Folio	18 × 12	50
Foolscap.	18 × 14	58
Half Royal	20 × 13	70
Crown	21 × 15	85
Half Imperial	22 × 18	106
Large Post	24 × 16	125
Demy.	24 × 18	130
Royal.	26 × 20	165
Elephant.	28 × 20	178
Double Crown	30 × 20	191
Imperial	32 × 22	224
Double Demy	36 × 24	294
Double Royal	40 × 28	394
Quad. Crown	40 × 30	422
Double Elephant	42 × 28	414
Double Imperial	43 × 33	516
Quad. Demy	48 × 36	648
Quad. Royal	54 × 40	834
Quad. Double Crown.	60 × 40	900
Quad. Imperial.	62 × 42	1016

Hints for Transferrers.—A correspondent of the "American Lithographer and Printer" makes the following valuable suggestions: "As a draughtsman and transferrer of considerable practice, may I be permitted a few words in favour of the roller in bringing up crayon and line work on stone? I always see that the transfer ink is of good quality, kept fresh, and never allowed to harden upon the roller. If the work is transferred to stone and appears weak, wash with clean water; then, if practicable, heat the stone slightly and allow it to lie without gum about two hours longer. This I have found always gave much better results than any manipulation with sponges could effect. Let any transferrer use a good magnifying glass and he will observe extremely minute fibres charged with the ink, which leave the sponge and adhere to the ink already on the stone. These fibres are unaffected with water; the acid burns them away, but leaves the ink adhering to and making larger the dots of the crayon drawing. Hence the drawing is made heavier and its delicacy of tone injured. A substitute for sponge may be had in a piece of fine woollen cloth, or, better still, a piece of uncoloured silk. These have not the same tacky qualities as the sponge, but after being used several times have to be cast aside, owing to the glazing of the ink upon them. Having tried them and several other methods, I always found the roller to be the quickest, cleanest, and in every respect to give the best results."

Litho-Bronze Printing.—It is at times required to print in bronze, at short notice, both sides of ball programmes and similar work. In ordinary procedure, one side would be allowed to dry before the other was printed, but in the case supposed there is no time for this. The secret is to employ drawing-paper or ivory cards, which are not very absorbent of ink. The stone having been made up to work both sides at once, a stiff ink is employed, and the cards printed and backed before the bronze is applied. This will be found quite effective and more simple than bronzing one side and then printing and bronzing the other. The second printing, in the latter case, would be sure to force the ink through the first applied bronze and necessitate re-bronzing, while in the mode we recommend there is sufficient ink left (despite the set-off taken from it) to hold the bronze, and consequently one bronzing is all that is required.

Hints for Lithographic Printers.—Lithographic

printing, to be successful, requires great experience and long practice, and practice cannot be taught; but we will give here a few theoretical instructions, which, when exactly observed, will give the printer the necessary practice very quickly, and we are sure many of our readers will thank us for this, because they will find here reasons for this or that which they see every day without knowing the why and wherefore of it. Of course, the perfectly posted practical man will say he knows all this; but there will be these few against the great mass who do not know it. We see printers in great number who do not know what is a good or what is a bad impression. It is a shame here to say this, but nevertheless it is the truth. A good impression is when a printer makes a clear, sharp, bright impression from a drawing or transfer as the work is on the stone. Some very good printers, and we know a few of them, make even a better impression than the work that is done by the artist or transferrer. Those who do not understand us will say this is impossible; but it is not. The first-class printer knows how to handle his roller in such a manner that he is able to deepen the colour on one part of the drawing and lighten it up on another. When you commence to print on a good drawing or a good transfer, use a regular medium ink, neither too stiff nor too thin, and observe the following rules :—Put on a pressure of about twelve pounds with your hands on the roller, the weight of the roller not included ; and observe that bearing heavier on the roller fills up the work, while bearing lighter on it takes the ink off the stone or drawing. A light pressure gives only a little ink to the work on the stone, and no pressure at all with the roller takes some ink away that has already been applied. Slow rolling overloads the work on the stone. Quick rolling takes the ink away, and takes the work off the stone. Slow rolling and heavy pressure on the roller give all the work more strength. In doing this, use only a little ink on your roller, and keep this in remembrance. When everything is in good order, this manipulation is the best. Quick rolling and light pressure on the roller sharpen filled-up work without any other help than that of the roller. If the work is very closely drawn, use a stiff ink. If the work is very weak on the stone, use a thinner ink, rolling slowly and with a medium pressure. In the summer-time, when the stones are warm, use only stiff ink ; thin ink in a warm temperature fills up the work. Cold stones are more dangerous yet. On a cold stone use thin ink, but even then the work works off the stone sometimes. When a stone dries too quickly, put a little glycerine into the damping water ; even a little salt would not hurt it. Transferrers should not do this, because this is only said for printers on hand or steam presses. In the foregoing instruction is given the whole theory of practical printing, containing more than many a printer learns in a lifetime.

Litho Decoration of Wood, etc.—The lithographic process in connection with the decoration of wood, iron, and other surfaces, varies somewhat in details, no two houses perhaps working exactly alike. The transfer paper is made by giving first two coats of starch composition, followed by one of thin gum, after which, so as to get a perfectly flat sur- face, the paper is rolled each way. Supposing the job to contain lettering, it would be drawn on the stone same way as read. Body colours are printed first, tints next, followed by bronzes, and lastly a surface of flake white covering every- thing up. The wood to be decorated receives a thin coat of transferring varnish, and before it is dry the print is thereon transferred. When the varnish is quite dry the paper is damped at back, and after a short time it will readily come off and leave the printing behind. A coat of fine varnish com- pletes the job. Metal plates are treated in a similar manner, but, to set the work, are baked before the final coat of varnish is given.

Fading of Red Lithographic Inks.—It is a matter of fact that red, especially cardinal lake, being an aniline colour, will fade, as will also the usual compositions of cardinal lake and vermilion. To obviate this difficulty, when a fine bright red is required, by mixing a little lemon yellow, say one- fiftieth part, with the cardinal lake and vermilion, a much finer and brighter red is produced, and one that will not fade.

Autography is the operation by which a writing or drawing is transferred from paper to stone, presenting not merely the means of abridging labour, but also that of re-verting the writings or drawings into the direction in which they were traced, whilst, if executed directly upon the stone, the impression given by it is *inverted*. Hence, a writing upon stone must be inverted from right to left to obtain direct impressions; but the art of writing thus is difficult, while, by means of autographic paper and the transfer, proofs are obtained in the same direction with the writing and drawing. *Autographic Ink* must be fatter and softer than that used directly on stone, so that, though dry upon the paper, it may preserve sufficient viscidity to stick to the stone when pressure is applied. This ink is made of white soap, one hundred parts; white wax, one hundred; mutton suet, thirty; shellac, fifty; mastic, fifty; lampblack, thirty or thirty-five; the whole mixed in the same manner as for lithographic ink.

An Improvement in Lithographic Damping Rollers has been made by a New York printer. Instead of the numerous rolls of felt, flannel, or other material now used to make up the desired thickness he interposes between the iron core and the outer covering a thick india-rubber tube, which renders the roller elastic, protects the metal from corrosion, and does away with many thicknesses of flannel or felt, at the same time keeping the roller in good shape, no matter what sized stone may be in use.

Lithographic Chalk.—Common or Castile soap, $1\frac{1}{2}$ oz.; tallow, 2 oz.; virgin wax, $2\frac{1}{2}$ oz.; shellac, 1 oz.; lampblack, 1 oz.

How to Wash Out a Job with Safety.—A drawing newly placed upon the stone, whether by transferring or drawing, may be easily injured by being washed out too cleanly with spirits of turpentine. The old composition of turpentine, oil, and gum-water, shaken up into a froth, was correct in principle and easy enough in practice, though

rarely to be found in use at the present day. Its only draw-
backs are that it requires a separate bottle to keep it in, and
necessitates the foresight to prepare it. As all lithographic
printers have the materials at hand, the same results may be
obtained by the addition of any kind of oil, in the ordinary
way of washing out. Proceed thus :—Roll in the work, if
the ink is dry, and then apply gum-water. Throw a few drops
of oil upon the wet stone, and then a few drops of turps ; then
rub it with a rag or flannel until the ink is dissolved. If it
be necessary to give it much rubbing, add a little thin gum-
water from time to time, to prevent the stone from getting
dry. The office of the materials may be thus stated : the
turps dissolves the ink, the oil preserves the necessary fatty
character of the drawing, and the gum-water keeps the rest
of the stone from contracting scum or tint. This is a safe
method, and in no way injurious.

Preparations for Zinc Plates.—There are few lithographic printers who have not understood that zinc plates may
be used as substitutes for stones in lithography, but the
majority of them have no knowledge of how they are prepared for printing. Lithographic stones are sensibly porous,
and are equally ready, when clean, to take up water or grease.
Zinc plates, on the contrary, are practically too dense to receive any but a superficial application of either grease or
water; hence it becomes necessary to produce upon them an
artificial coating having a distinct affinity for gum and water,
and a quality of resisting the encroachment of the fatty ink.
This desirable quality is best secured by a solution of nut
galls, applied with a brush, followed by gum-water, or,
according to the practice of some workers, the gum may be
mixed with the galls. Make a decoction of the best Aleppo
galls by steeping 4 oz. in 3 quarts of water for 24 hours, and
then boil up and strain. The nuts should be first broken up
small. Then take—

Decoction of Galls .... $\frac{3}{4}$ pint.
Solution of Gum (thickness of cream)......... $\frac{1}{4}$ „
Solution of Phosphoric Acid .... 3 drachms.

Another Etching Solution, much used among German printers, may be made as follows:—

Nut Galls, bruised	1¼ oz.
Water	! pint.
Nitric Acid	2 drachms.
Acetic Acid	4 drops.

The galls are boiled in the water until it is reduced to one-third its bulk. To the strained liquor the acids are added.

Both these recipes are doubtless very good. Their value lies in the nut-gall solution, however, and that may quite effectually be used alone. The gummed plate must be dried quickly by artificial heat; the ink should be applied by the roller while the gum is dry, and when fully blacked the water should be sprinkled over it, and the rolling continued, with occasional sprinkling, until the plate becomes clean. It is then ready for printing in the ordinary manner of lithography. A lay-mark may be made with a silver coin.

Preserving Drawings on Stone after Printing.—

It is too often the practice in lithographic printing offices to take but little notice of the stone when the first order has been executed, but if there be only a remote chance of its being required again means ought to be taken, as far as possible, to insure that the stone be in fair printing condition when another edition is called for. The stone may not be wanted again for months, or even years, but the ordinary ink may become so dry in a few weeks as to become insoluble in turpentine, and to have lost its power of resisting the adhesion to it of water. Hence the necessity of preparing it in some manner that will permit of the removal of the ink by turpentine, so that the stone will be in a similar condition as when first printed from. Drawings may be preserved by the use of the following ink, to which some recommend the addition of soap; this, however, is not only quite unnecessary, but may prove really mischievous:—

Ordinary ink, as bought from the maker...	2 oz.
Tallow	2 ,,
Beeswax	4 ,,

The tallow and wax are to be melted, and the printing-ink added a little at a time until dissolved. When about to be used, a small quantity must be ground with turpentine until of the consistency of ordinary printing-ink. Wash out the drawing with turpentine in the ordinary way, and roll-in with the above ink until the drawing shows clearly, using a small quantity of gum-water on the stone to keep it quite clean. Set the stone aside for a few hours until the turpentine has quite evaporated, and then gum-in.

Another Method.—Roll in the stone with ordinary printing-ink. Dust with powdered resin and allow time for the ink and resin to incorporate and become hard. Take a spoilt impression of the job, and gum over the back of it with gum-water, lay its gummed side to the stone, and pull through the press. Gumming the paper instead of the stone will more effectually exclude the air, and thus prevent " oxidation " of the ink, for which " drying " is only another name. If the stone is to be laid by in a very dry place, the addition of a little glycerine to the gum will prevent its cracking. It is better than sugar, molasses, etc.

How to use Lithophine.—Lithophine is a new material, of American invention, said to make the drawing more permanent upon the stone, and may be applied to work partially worn out, or to transfers newly laid down when they show a little tendency to weakness. It may also be used upon any work as a precaution in preventing deterioration. We can recommend the following method of using it, in preference to that given in the instructions sold with it, though it does not differ from it in principle:—Gum in the work with gum that is not very thin, and dab it over with a cloth to prevent it overlying the ink. Allow the gum to dry, and then wash out with lithophine, which must be well shaken up, and allow it to evaporate on the stone, which it does rapidly. The gum must be then washed off with plenty of water, and the job rolled-up. Treated thus, it has the maximum effect, consistent with freedom from scum or tint, upon the stone, to which it is somewhat liable if carelessly used.

How to use Soap as a "doctor."—It sometimes
happens that a drawing by accident gets over-etched, or
some acid gets dropped upon it, so that the work so affected
will not receive the ink from the roller. The use of lithophine
in this instance may or may not be instrumental in restoring
it. Rubbing-up with gum and retransfer ink, or ink mixed
with palm oil or gold size, will perhaps only temporarily
improve it. When the roller is applied, the ink comes away
from the weak place because the ink fit for rolling-in the job
is stronger than that applied with the rubbing-up rag. Not
only so, but the rubbed-in ink has not penetrated to the
stone sufficiently to have a firm hold. Strong litho-writing
ink is sometimes used with the rag in a similar way, and in
extreme cases a strong solution of soap is carefully applied
with gum to keep it off the clean part of the stone. The
mode of applying soap is wrong in principle; for being
soluble and used with gum-water its tendency to penetrate
the stone is almost greater where the work is not than where
it is, because it is like applying soapy water to a place
partially greasy. It is possible, however, to use the soap
remedy in a way at once more scientific and effectual.
Having temporarily restored the work by rubbing it up in
gum with ink and oil, gum-in carefully, and dab it over with
the damping cloth in such a manner that the gum is left on
the clean stone, but not on the lines of the drawing. Allow
the gum to dry hard, so that it does not feel clammy to the
fingers. Wash out the ink clean with turpentine, with the
gum still dry upon the stone. Now take a piece of *dry* soap
and rub it well into the weak place. It will not dissolve the
gum, which therefore protects the stone from its action,
excepting where it has been inked. Remove the superfluous
soap by rubbing it with a rag charged with oil. Dissolve
the gum and roll-up, when most likely the work will be
restored. If not quite so, repeat the operation. This method
is very effectual if properly used, and produces no scum to
be cleaned away.

A Lithographic "Wrinkle."—The "American Lithographer" has awarded a prize of one hundred dollars for the

following "wrinkle," the sender of which thus describes his
novel and simple discovery :—" All our lithographers, printers,
and pressmen who have ever printed from zinc in the lithographic press are aware that, despite the greatest care, the
work upon the zinc plates will occasionally smear or tint.
In a word, the printer is unable to keep the work clean upon
the plates. Almost by accident a remedy has been discovered.
One day, while working overtime, I was printing from zinc.
My damping water had been used up, and I was too lazy to
go for more.   Near at hand was some tea which I had made
for drinking, but had found it too strong for my taste.   It
was without milk or sugar.   Because of pure laziness, I used
this strong tea for damping rather than take the trouble to
go to the hydrant for water.   To my surprise, after using the
tea, the plate, which before had smeared very much, became
gradually cleaned, until the transfer was as nice as could be
desired and as clean as when the plate was first made.   Then
my fears began to stir lest through continuing the tea for
damping the work would gradually get too weak, or, as we
say, 'too sharp.'   I therefore very closely watched the work,
but my fears proved groundless.   From that time I have
used only tea for damping and have had no trouble, but
always made good work.   I have also found it useful in
printing from stone.   My discovery I have heretofore kept
sacred as a secret.   As to the reason for the effect of which
I have spoken, I cannot speak.   You probably know more of
the chemical properties and action of the tea, and probably
can account for the matter."   The editor adds : "Tea contains
from six to twelve per cent. of tannic acid ; it is therefore
very probable that the above writer is correct in all his
statements."

Zinc Plates as Substitutes for Lithographic

Stones are now in common use, and an ingenious and cheap
method of producing a stony surface upon a metal plate, to be
used in a lithographic press instead of the ordinary stone, has
recently been patented.   Slacked lime is added to a water-
bath, which is afterwards treated with carbonic acid.   A
saturated solution of bicarbonate of lime is thus produced,

which can be drawn off as a clear liquid. A carefully cleaned metal plate is moistened with this liquid by a spray apparatus, and then dried by heat. The operation is repeated until a deposit of limestone firmly adheres to the plate, when it is ready for receiving the lithographic ink.

The " Glazing " of Litho Rollers is accomplished by rolling them up in a colour, such as orange-lead, that dries quickly either with or without driers, and varnish.

Lithotint.—A process of "lithotint" was patented by Hullmandel in 1840. The drawings were made in washes upon the stone, covered with a solution of resin, pitch, and spirits of wine, and when dry the stone was etched with nitric acid in gum-water.

New Process of Gold and Silver Printing in Lithography.—By this process the separate printing of bronze can be saved, without new machinery, without any extra expense, and without any other workmen than the practical artists and printers who have the necessary brains to understand a plain thing. The process is only a matter of knowledge, requiring no special machinery or materials. The *modus operandi* is as follows:—Suppose that a show-card, label, or anything similar, is to be printed in colours, say, for example, silver, gold, yellow, red, blue, black, and flesh colour ; these seven colours can be done in five printings. The artist finishes the yellow, red, blue, flesh and black plates in the usual way ; but before etching he adds the gold plate to both the blue and the original plates. No matter if this comes in contact with the other work ; remember that the blue and gold are to be done at one printing, and also red and gold. Add to the original yellow and flesh-colour drawings the same work that would be necessary for the silver printing ; for the same manipulations have to be made as with the red and blue and gold plates. Leave the black as originally drawn. In printing do not use any dryer in the colour. Print the yellow first, in the usual way, and let the sheet get perfectly dry : the time will vary according to the ink and paper used. Now print the flesh colour with, in winter, mostly No. 2 varnish; in summer, mixed with No. 3 and 0. Let the impressions lie, according to the paper, from one to five hours, and then run them, for the silver bronzing, through the bronzing machine, or bronze them by hand when there is no machine. You will now find that where the flesh colour is done on the clean paper the printing is dry and will refuse to take the bronze, and where the flesh colour is printed over the yellow the bronze will stick solid, even if done on the second day. It is only required that the bronzing shall be done at the right time ; but this is not a matter of so much anxiety as one might suppose. As in all printing, it requires only ordinary care. The impressions have to be dusted off in the usual manner, and then the third printing should be the blue. Let the ink dry again, and then print the red. The same care and manipulation should be had over again as stated with regard to the silver. After the gold is dusted, print the black. This is the full process, given so plainly that we think every artist and printer can understand it. The advantages are as follows : two printings are saved, and the customer may be charged with the full seven printings, as this is an advantage to the lithographer from a knowledge of a special thing, and has nothing to do with one's obligations towards the customer. When we take into consideration that bronze is desired in nearly all lithographic work—that there shall be gold or silver seen on it, as the customer says—and when we can do it, as now, without incurring any extra expense, it should be easy to understand that this is an advantage which ought not to be undervalued, apart from the fact that bronze printing is generally charged the price of two printings.

Tympans for Litho Presses.—Tympan leather is largely used by lithographic printers. The hides are sometimes dressed whole, being usually hides set apart for that purpose ; shaved hides may be used if they run stout. The hides should average from twenty-five to thirty-five pounds each. A writer in a leather-trade journal says that, in practice, he has found it the best way to cut the bellies off in the rough state, that is to say, before they are soaked down. Great care must be taken in selecting the hides for tympans, as they must be quite free from flaws or cuts on either side. It is better to select good grown hides, as this gives the advantage of cutting large-sized tympans, and the veins are easily got out without reducing the substance of the leather. From a well-grown hide tympans can be cut close up to the neck. When the hides are dressed whole, that is, with the bellies on, there is not the possibility of getting the stretch out, which is very necessary.

Another Method to Protect Drawings on Stone.

—The protection of drawings on stone is a question in all lithographic houses. The rule is to roll the stone up with colour and gum it; but, excellent as this plan is, the gum is sometimes liable to scale away, leaving the stone exposed to damage. A Paris paper gives a new plan to the same end, which has been applied with great success. The preservative compound is composed of the following ingredients:—150 parts spermaceti, 140 parts Burgundy pitch, 90 parts olive oil, 50 parts white wax, 30 parts Venetian turpentine. These to be melted together. The composition is applied to the stone with a roller. It covers the stone and protects it in all temperatures, even if exposed to the weather out of doors.

Covering Machine Damping Rollers.

—The best material for this purpose is known among tailors and drapers as *moleskin*. It is a thick twill cotton material, about one-sixteenth of an inch in thickness, commonly in use for clothing by engineers and other mechanics, and valued for its washing qualities. It may be got of almost any draper in the country, or of the warehousemen in large towns who supply tailors. For the purpose under consideration, it should be chosen of white or light stone-colour. Lay a piece of it out upon a smooth board, and with a sharp knife and straight-edge make a clean even cut across the piece, as one may be sure that the draper has not cut it straight enough with his scissors. Take a narrow piece thus cut off and accurately

measure the roller round its felt or flannel undercovering.
Measure it at each end, so as to be sure of making the new
cover fit with accuracy.   Then cut a strip the proper size
and length for the roller.   Now, with a stout *waxed* thread,
proceed to sew it on the roller, putting the needle in on the
under side of the edge and pulling it out on the upper side,
taking hold of about one-eighth of an inch of the stuff at each
stitch, carefully pulling the thread in such a manner as to
make the edges meet neatly, but not to overlap.   After sewing an inch or two, it will be seen if the cover is the proper
size.   If too small, cut another ; if too large, take it off and
cut a strip off it.   When sewn on, secure the ends something
like ordinary roller-ends.   Previous to being put in the
machine, they should be thoroughly wetted with boiling
water, as, when the moleskin is quite new, it has a tendency
to repel cold water, while it takes hot water readily.   One of
these rollers will damp nearly as effectually as two covered
with canvas, and will last many times as long.

Cleaning Lithographic Rollers.—If the ink has dried
on hard, soak the roller in turps for a day or two, then scrape
off the ink with a knife, wash with turps and wipe with a
clean rag.   A little salt used in conjunction with the turps
will assist the operation.

ENGRAVING AND PROCESS WORK.

THE Oxidation of Zinco Blocks.—A correspondent of the "Austrian Printers' Gazette," says that with zincotypes the greatest possible cleanliness is important, as oxidation occurs rapidly in this kind of engraving. Zinc oxidizes quickly when exposed to the air or to alkaline liquids ; when once formed, it freely develops under the influence of the soda potash. He recommends the use of benzine or spirits of turpentine, drying with a rag, and then placing the zincotype in a drawer. When washing formes, dirty water or such as has already been used is often taken. This latter always contains potash, petroleum, spirits, and dirt. This is another cause of rusting. Another habit is no less injurious. When the potash does not act quickly enough, some washers are accustomed to pour spirits of turpentine on the brush or on a rag and to rub the rebellious spots, without taking the precaution, however, to use a little potash and clean water afterwards. It has been ascertained by chemists that spirit of turpentine, especially when old, absorbs oxygen from the air and ozonizes it—that is, it transforms it into an active and positive oxide that acts very energetically—so that any spirit remaining on a forme not only favours, but actually excites the development of the oxide on the type.

Collotype.—This process is a method of printing from gelatine, glue, and similar "colloid" materials, to which a pigment is added. The word is derived from the Greek κύλλος, glue. The process is called in French *heliotype* or *phototypie*, in Germany *lichtdruck*.

Substitute for Sulphuric Acid in Etching.—The use of sulphuric acid for etching on zinc and copper is connected with so many disadvantages, and so injurious to health, that a substitute has long been sought for. Herr Krätzer, Leipzig, recommends the following mixture for the purpose. Forty-five grammes of fine powdered gall-nuts are put in 600 grammes of water, and the mixture boiled down to one-third. It is then filtered through a linen or felt filter, and three drops of concentrated nitric acid, as well as from four to five drops of muriatic acid, are added to the filtrate. This mixture is suitable especially for zincography. For fine work it must be somewhat diluted, and left only for a few minutes on the zinc plate. The latter is then carefully rinsed with water, and covered with a fresh and diluted solution of gum arabic. For treating copperplates, 150 grammes of fuming hydrochloric acid are diluted with 1,050 grammes of water, and a boiling solution of 30 grammes of chlorate of potassium in 300 grammes of water is added. If weaker parts are to be etched, the solution may be diluted still further with from 1,050 to 2,000 grammes of water, whilst the deeper shades are produced by the addition of a stronger liquid or by exposing the plate longer to the action of the solution.

New Engraving Machine.—With the invention of the pantograph many easy methods of drawing and reproducing linework became possible, and a new adaptation is in the form of an engraving machine, capable of copying on metals, ivory, and glass, whether they be of flat, bevelled, or cylindrical surfaces. In this machine the style is guided by the right hand of the operator, while the left hand regulates the depth of the cut by means of a milled nut, which enables the operator to gauge the depth or lightness of the cut, or keep it constant at will. The cutter runs at a high speed, and the section of the cut may be of any form, moulded, semicircular, bevelled, rectangular, or even dovetailed. For purposes of engraving or finishing brasswork a simple, flat double-edged drill gives a clean, brilliant cut free from burr, which may be either left bright or coloured by suitable solutions as wanted. When it is desired to fill the cut with wax, a long pointed drill is used,

which gives a cut with steep sides, and leaves the bottom of
the cut sufficiently rough to hold the wax. Lines are worked
by a simple point tool in the cutter spindle, the style being
moved along fixed straight guides, and all lettering may be
done at the same setting of the machine. The machine
was originally invented by a firm in Leicester, for engraving figures and letters on lenses and other scientific
instruments, but its undoubted capacity has been adapted to
a great variety of work, including name-plate engraving, embossing seals, dies and moulds, marking and numbering instruments and tools, dividing and engraving dials, scales for
gauging and measuring instruments. The machine is also
equal to the performance of engine-turning and profile work,
and by the use of suitable milling tools, in conjunction with
the dividing apparatus, wheel-cutting and similar work may
be executed. The engraving, of course, is not of the same size
as the pattern, and the drawing may be varied from one-fourth
to one-sixteenth of that of the copy.

Steel-Facing Photogravure Plates.—The number
of good impressions that can be taken from a copperplate are
necessarily limited, and the vexatious process of multiplying
the original by the galvano-plastic process has been superseded by the process of steel-facing the plates. When a
copperplate is placed on the cathode, suspended in a solution
of sesquichloride of iron and subjected to the action of the
galvanic current, it will in a short time be covered with a
delicate and lustrous cuticle of iron hard as steel, from which
many thousands of impressions may be taken, equal in every
respect to copper or steel-faced plates. The process is
carried on in a peculiarly arranged dark trough with a three-
cell zinc carbon battery, and the electrodes are placed verti-
cally. The iron solution is made suitable to the current
itself. One part of chloride of ammonium is dissolved in ten
parts of water, and in this solution are placed iron plates as
cathode and anode. By chemical action the chlorine unites
with the iron of the anode, forming sesquichloride, which
remains dissolved in the bath, which within a day or two will
assume a greenish colour, and owing to the formation of

hydrate of oxide of iron, consequent on its contact with atmospheric air, its surface acquires a red scum, and a metallic mirror will make its appearance on the cathode. The bath is now sufficiently saturated, and in place of the iron cathode, the copperplate, having been previously cleaned in a solution of caustic potash, rinsed, and any adhering alkali neutralized with weak sulphuric acid and again washed and dried, is substituted. The steel cuticle can, when worn by long usage, be easily removed by laying the plates in sulphuric acid diluted so that the copper will not be affected by it. The tendency of the acid is to detach the steel from the copperplate.

Photo-Etching on Wood.

—A Russian inventor has, by a very ingenious process, extended the art of photo-etching to boxwood blocks. The block is boiled in two separate solutions, by which the pores are filled with insoluble carbonate of copper. It is then polished on the surface, coated with a solution of asphalt on the back and sides, and the face is covered with a gelatine film. The photograph is then taken on the block, and the soluble portions of the gelatine having been washed out, the remaining surface is coated with asphalt. The block is then placed in strong nitric acid for an hour, and afterwards for the same period in sulphuric acid, which changes the unprotected portions into nitro-cellulose.

New Process of Engraving.

—A new method of producing letterpress engravings has recently been perfected. Briefly described, the process is somewhat as follows :—A flat, highly polished steel plate is coloured black and covered with a soft white composition. By means of tools made for the purpose—somewhat like those used for wax engraving —the design is drawn through the composition, exposing the dark background. When finished the drawing appears exactly as it would on paper, the only difference being that the lines are sunk *through* the white surface. The intaglio or matrix thus formed can be placed in an ordinary stereotype casting-box and cast. The hot metal will run into every line and produce a perfect reverse or relief-line engraving. If the composition has been scraped to a proper thickness no routing will be necessary, and the cut will have as much relief and as sharp lines as one made by any other process. The composition is soft, and the sketching may be done with a lead pencil. Tracings can be made on it, or the artist can work directly on the plate; ruling can be done with either a T square or a ruling machine. Being strictly a graphic process no experience as an engraver is necessary. Of course, the operator cannot make a better cut than he could draw on paper, but with a little practice he can do good work on the plates. No chemicals or photographic apparatus are requisite. Not only is there a saving of space, but there is a freedom from disagreeable smells and dirty accumulations. The material is prepared for the artist's use, and when the drawing is finished, the stereotyper's aid is alone required to have a perfect printing block. If there is a stereotyper at a convenient distance the artist will need only a set of tools and as many plates or parts of plates as he wishes to make engravings. The saving of time is enormous. It takes but little more time to draw on the plates than to draw on paper, and the cut can be made ready for the press in half an hour, when the drawing is finished. In all other processes fast work is a matter of hours. The drawing is right-handed, not reversed as required for every other process. Reversal often alters the whole aspect of the cut. Having his work before him just as it will print, the artist can do it much better, and his idea cannot be altered or spoiled by an incompetent engraver. The rapidity with which cuts can be produced has made newspaper illustration more practicable. While peculiarly adapted to rapid work, fine commercial cuts can be made by the process. The most delicate lines and the finest cross-hatching or shading can be reproduced.

Etching of Iron.—Use the following mixture: Hydrochloric acid, one pint: water, one pint; concentrated solution of antimonious chloride, one drop. The last ingredient is added to prevent rusting of the etched parts. Soft and fine-grained metal is more easily acted on than any other sorts.

Etching by Process.—It is important for zinc-etchers to know that no more acid solution is necessary than scarcely enough to cover the plate. By constantly moving this acid solution over the plate, which is best effected by having the containing vessel in a swinging position, the air can all the time strike the plate. The acid must never cover the plate ; it must only pass over it from the motion of the plate itself or the containing vessel. This makes a quicker and more even etching than by the old plan.

Recipe for Engraving on Glass.—Cover the surface of a sheet of glass with a concentrated solution of nitrate of potash, by simply placing the glass flat on a table or in a shallow pan, and pouring the solution upon it. Then along the edges of the sheet place a platinum wire, keeping it immersed in the solution, and place it in communication with one of the poles of a secondary battery. Having done this, let another fine platinum wire be joined to the other pole, and incased in an isolating substance save at its point. With this wire trace on the glass the design required ; a luminous streak will appear everywhere the wire touches, and however quickly it may be moved the design will be cleanly engraved on the glass. If the drawing or writing be done slowly the lines will be engraved more deeply. Their width depends on the diameter of the wire at its end : if it be reduced to a fine point, the work may be performed with great exactitude. The metallic thread conducting the electric current thus becomes transformed into a special graver for glass, and in spite of the hardness of the substance, the operation requires no effort, for the corroding force is furnished by the action of the current upon the saline solution. Either pole may be used for engraving, but it requires a weaker current to engrave with the negative pole. If instead of a plane surface it is desired to engrave on a curved one, the same result may be obtained by thickening the solution with a gummy substance, in order to make it adhere to the glass, or by turning the object in a basin containing the solution, so that a freshly wetted surface may be continually presented to the operator. These remarkable results have been obtained by means of secondary batteries, but for continuous work any other source of electricity may be made use of, provided it has volume and intensity enough. Thus a pile of a good many Bunsen's cells, or a gramme machine, or even an electro-magnetic machine, with alternate position and negative currents, will do.

Etching Fluid for Steel.—The following is highly praised for being an excellent etching fluid for steel: Mix one ounce of sulphate of copper, one half-ounce of alum, and one-half a teaspoonful of salt reduced to powder, with one gill of vinegar and twenty drops of nitric acid. This fluid may be used for either eating deeply into the metal or for imparting a beautiful frosted appearance to the surface, according to the time it is allowed to act. Cover the parts necessary to be protected from its influence with beeswax, tallow, or some other similar substance.

Process for Transforming Zincos into Engraved Plates.—The process in question, says " L'Imprimerie," permits of the rapid transformation of ordinary zincos into engraved plates, and even into copper plates; that is to say, a zincographic plate, the bitumen of which is frequently destroyed in working, may, by this process, be replaced by an indestructible engraved plate, capable of furnishing an unlimited number of copies. The details of the operation, which will, doubtless, interest our readers, have been communicated by Captain Bény to the Société de Photographie, and are briefly as follow:—Take an ordinary zincographic plate, the bitumen of which is already inked, pass it quickly through a solution of nitric acid, clean it with a soft brush, wash thoroughly, and plunge it into a copper bath, where it should be allowed to remain five minutes. The zinc becomes coated with copper, which adheres to every part left unprotected by the bitumen. The heavier parts of the image contract, and are even rendered finer by the action of the copper bath. When the whole surface of the zinc has assumed a fine red, the plate is withdrawn and placed in a basin containing a little pure benzine. The bitumen is now removed with a brush from the surface of the plate, and the drawing shows a bright white on a sombre, copper-red ground. When the zinc is freed from the varnish which remained on the drawing, the surface is thoroughly washed in order to remove every trace of benzine, and afterwards rubbed with a rag or a sponge in a tub of water. This done, the plate is placed, without being wiped, in the nitric acid bath previously used. It will then be seen that the naked zinc of the image is alone bitten, the coppered surface having resisted the action of the acid. At the end of fifty to sixty seconds a fine plate will have been produced, without either a drawing or a positive *cliché*. The making ready and inking are done in the ordinary manner. A more prolonged biting, demanding, besides, on the surface of the zinc, a thicker coating of copper, strengthened, if need be, at the battery, will give zincos resembling copper plate. In this case, in order that the various parts of the bitumen may be preserved intact, they should be thoroughly inked before coating the naked zinc. As the copper afterwards resists the acid during the process of engraving, all defects in the zinc will be obviated. This process is, therefore, valuable for obtaining thickly-coppered zinc plates at a much cheaper rate than those made entirely of copper.

Typographic plates may also be obtained by this process. For this purpose positive *clichés*, or very dark drawings on translucent paper, are used, the manipulations being the same as those before indicated.

Engraving with Mercury and its Salts.—It is known that, when mercury is deposited on a metal, fatty lithographic ink will not "take" upon it when an inking· roller is passed over it, and that the black adheres to the untouched parts of the metal. If a well-polished and clean piece of zinc be taken, and a design be traced thereon with mercury, the design will appear in brilliant white upon the grey background of the zinc. After tracing the design, an intaglio plate can be obtained by plunging the plate, without being coated with varnish, into a bath containing 100 parts of water, and 2 parts, at least, of nitric acid. The action of the acid is very rapid, and for a long time only attacks the parts touched by the mercury. When deep enough, it can be used for lithographic work. If, instead of nitric, hydrochloric acid is used, the contrary effect takes place. The unaffected zinc is strongly attacked, and the traces of the mercury give a relief plate which can be used for ordinary typographical work. If the operator does not wish to draw upon zinc, the design can be traced upon paper with a salt of mercury. The sheet of paper being then applied for two hours to a plate of zinc, the drawing is sharply reproduced, in white lines of amalgam, on the grey surface of the metal, just as if it had been traced directly. The same result is obtained if the design is traced upon paper with a sticky substance (ink containing gum or sugar), and if it is dusted over a mercury salt in fine powder. On dusting off the surplus and applying the sheet containing the design to a plate of metal, the same result is obtained. The same result is obtained if a newly-printed proof is used, and is dusted with mercury salt while the ink is still wet and sticky. All the lines thus reproduced are chemically engraved, as has been described above. The same results are obtained by dusting with mercury salts a photographic carbon print containing a gummy substance, and the effect of half tints is even secured. Biniodide of mercury is the salt to use.

A New Process of Illustration.—An effective process of reproducing wood-cuts, engravings, and other printed matter direct from the original, it can readily be seen, must be an exceedingly valuable innovation. The manifold reproduction of such matter has been attempted a number of times, with different degrees of success. It may, therefore, be interesting to describe a method that has been in use on the continent. The object to be reproduced is first well saturated with a protective solution composed of glycerine, tannic acid and water, then well rinsed with water, and the parts to be reproduced treated with autographic ink by means of a suitable inking roller, or in any other appropriate manner. In consequence of the previous treatment of the original with the protective preparation, ink will only adhere to those parts covered by printers' ink, whereas those parts left free will not take up any quantity of ink. The positive so produced or written, or drawn in autographic ink, is now laid on a brightly polished metallic plate or lithographic stone, and repeatedly passed through the rollers of a press of peculiar and adapted construction, so as to transfer the autographic ink of the positive to the plate or stone, thus producing a negative copy. The protective mixture of glycerine, tannic acid and water is then applied, and the plate or stone provided with ink by means of the inking roller. It is preferable to ink the plate or stone by passing it repeatedly between an upper inking roller and middle roller with which the press is provided. The paper, or other matter to which the ink adhering to the negative is to be transferred, is now placed over the negative and passed between the pressure rollers of a suitable press, whereby the exact and true copy of the original will be found reproduced. After the paper has been removed, the plate or stone is again inked; paper is then applied to the surface of the same, and a second copy produced in like manner as aforesaid, and so on, so that many hundreds of reproductions or copies can be produced, which are in all respects, it is claimed, perfect facsimiles of the original engraving, or printed, drawn, or written matter.

Method of Galvano-Engraving.—The new art of engraving metallic plates, to be used for printing and ornamental purposes, and styled galvano-engraving, appears to possess some peculiar advantages as a method of photo-engraving metallic plates. Thus, to make an engraving, there are suitable metallic plates prepared, which have the smoothness and polish of glass, and having obtained a photographic negative on a glass plate of the subject to be engraved, the operator next covers one of the polished plates with a bichromated gelatine film, places the photographic negative upon it, and exposes it to the light. The action of the latter renders the gelatine insoluble, so that when the negative is removed, and the gelatine plate washed, all the gelatine on the surface of the plate is removed except the duplicate of the lines of the photograph, these remaining in relief. The proof is placed for some hours in a damp place, where the lines are brought up in relief, and after the proof has been coated with plumbago, it is applied to a metal alloy placed in a special vessel; the alloy is then subjected to an ordinary pressure, and, on cooling, produces a hollow metallic plate ready to be printed. The fusible alloy employed consists of bismuth, tin, lead, and mercury, in proportions according to the degree of hardness desired. The vessel for containing the metal has a bottom formed of a smooth, strong metallic plate; into this vessel the liquid metal is poured, the gelatine proof immediately applied on the metal, and, the whole being covered by a second smooth metallic plate, which closes the vessel, momentary pressure is applied.

Hints on Drawing for Process Blocks.—All pen-and-ink sketches for this purpose should be pure " black and white." Freshly-made Indian ink should be used on thin white and smooth card, the Bristol boards obtainable from most artists' colourmen being well adapted for this purpose. The lines should be firm and distinct, the depth of light and shade being obtained by thick and thin lines respectively; the distance between the lines also helping this effect. The drawing should be made larger than the size intended for the block; from one and a half times to double the size is a good rule. A smaller amount of work thrown into the sketch is often more effective than an excess of pen-and-ink work. Care and attention to these details, with practice, will soon enable anyone with a knowledge of drawing to overcome any difficulties which might otherwise be encountered.

Process Half-toned Blocks.—A new process of photography applied to type printing, which for the first time enables the half-tones of a photograph to be reproduced at letterpress, is about to be introduced for practical working in this country. Its discovery and successful operation are due to Herr Meisenbach, of Munich, who has been employed for six years in perfecting his process. It is not restricted to illustrations, but is equally applicable to reproduction of type. The value of the invention is fully recognized in Germany, where six of the principal journals have been illustrated by it, as well as other works. In Paris a special number of " L'Illustration " was devoted to the reproduction of the chief pictures of the Salon by the process. It is also largely used for commer-

cial work—such as business catalogues, etc.—in most of the
Continental States. At a push the necessary type-printing
blocks can be produced within six hours, but always within
twenty-four, either by the sun or by the electric light. They
can be made at what is comparatively a small cost, and are so
enduring that not less than 160,000 impressions have already
been taken from one, and the block is still in use. The process
is being introduced in this country by a limited liability
company.

Photographing on Wood.—The method here given is
taken from Dr. J. M. Eder's " Year Book." Volkmer's boiled
arrowroot paste : One part of arrowroot to 100 parts of water,
mixed thoroughly with finely pulverized flowers of zinc, is
spread thinly and evenly over the picture side of the block,
either with the ball of the hand or a camel's-hair brush, and
dried for two hours. The block is then covered with equal
parts of albumen and water, to which a little chloride of ammonium has been added, and dried again for two hours.
Before sensitizing, the sides of the block are covered with
beeswax dissolved in benzine. Silvering is done by merely
touching a ten per cent. silver solution for 30 seconds. The
superfluous solution is allowed to drain, the block blotted off,
and dried. Printing is done with the aid of Vogel's photo-
meter, and completed when Nos. 12 and 14 have been reached.
The picture can be toned in the ordinary way, but should be
fixed in a very strong hypo bath.

Another Recipe from the same Source.—Mix pure
albumen with white lead and a little chloride of ammonium,
and distribute a few drops of the mixture over the surface of
the block. When perfectly dry sensitizing is done with an eight
per cent. silver solution for two minutes, and when dry the
block is fumed for 20 minutes in the vapours of ammonia and
then printed upon. After a slight washing the picture is toned
and fixed in the combined gold and hypo bath. This method
has been practised with invariably good results that are satisfactory to the engraver.

To Preserve Zincographic Etchings, etc., from
the oxidizing influences of the air, a highly rectified oil, free
from any acid, will do good service when laid on the plates
with a light brush. A still better means of protection is the
following:—The zinc plate is first washed with turpentine, then
with a solution of potassium, after which pure water is freely
applied to it to take away any remnants of dirt or ink. This
done, the plate is quickly dried on a heated iron plate, and
when still warm, a roller covered with the ink commonly used
by zincographers is passed over it, so as to apply a layer of
ink on all the raised parts of the drawing, and to give them a
cover for protection against atmospheric influences : or a layer
of asphaltum which has been pulverized and diluted in spirits
of turpentine will answer the same purpose. It may also be
applied with the same effect to all electros. Hot water or lye
will take it off immediately when the plates or electros are to
be used again.

Leimtypie.—The process consists in the employment of
chromated gelatinous sheets, upon which the subject is transferred. The development is very rapid, not occupying more
than from two to four minutes. It is produced by the
application of a liquid which has the property of dissolving the
glue not covered with work, the plate being brushed either by
hand or machinery. Plates with larger whites take about four
minutes, in order to get sufficient depth, and, should it be
necessary, the whites can be cut away as in the zinc plates.
When developed, the glue or gelatine sheet is fastened on to
a zinc plate in such a manner that it is impossible to separate
them. This is a special feature of the invention. It is claimed
that the entire block is finished sooner and costs less than the
negative that is needed for zinco blocks. The inventor claims
that Leimtypie plates render all the details of the original
much more sharply than zinco blocks, because with the
Leimtypie a direct transfer on to the equal surface of the glue
takes place, and this reproduces with mathematical accuracy
all the details of the negative. On the other hand, there is
always some impurity in the manufacture of zinc which

R

renders the development unequal. The copy of the picture takes place through albumen or some other medium, but never through zinc itself, and therefore many details, as well as the character of the work, are lost.

To Engrave on Steel.—A box, containing powdered cupri sulph., and labelled "The Powder," is required, then dissolve some of the powder in a small quantity of water ; rub the surface of the steel over with a piece of wetted soap, so as to cover it with a thin coating; then dip the point of a pencil into the solution, and with it write or draw the required design on the steel. After a few minutes wash, and the steel will be found to be beautifully and permanently engraved.

Fichtner's Process for making printing plates by photographing directly upon zinc, copper, or steel-plate is as follows :—A sensitive varnish is made by dissolving 5 parts of Syric asphaltum in 90 parts of benzine, adding 10 parts of oil of lavender and then filtering. A great deal depends upon the quality of the asphaltum. Select those pieces which do not melt at a temperature of 190° Fahr. and are difficult to dissolve in turpentine. The benzine must have been purified and perfectly free from the admixture of water. The same holds true of the oil. A perfectly clean zinc plate is coated with the sensitive varnish, exactly as the photographer would use collodion. The plate is rested next upon a level surface, in order to secure an even distribution of the varnish, and then dried in the dark chamber. After this the photographic negative is placed in a copying frame and upon it the prepared zinc plate, the two pressed closely together and exposed for a period of 25 to 30 minutes in the sun, or three to four hours in scattered light. The duration of exposure, of course, depends on circumstances, and has to be determined by trial. The photographic negative should have been transferred from the glass plate to gelatine to prevent the reversing of the picture. When exposed for a sufficient time, and after being taken out of the frame, no print should be visible upon the plate.

Rectified petroleum, to which one-sixth part of benzine has been added, is now poured over the plate until the drawing appears in the colour of the sensitive varnish and the open spaces show the clean metal. After this, wash in plenty of water and dry by the aid of a flame. The etching may hereafter be done in the usual manner with diluted nitric acid. *

Wood Letter.—Pear-tree and sycamore woods are chiefly used for cutting small wood-letter on, pine or deal for larger sizes and blocks. If pine or deal blocks are varnished and polished they will not absorb the ink and will print much cleaner. The timber of the pear-tree is of a yellow colour, and very firm and solid. The blocks with which the designs for floor-cloths are printed are made from pear-wood. When dyed black it can scarcely be distinguished from ebony.

Map-Engraving Wax.—Four ounces linseed oil, half ounce each of gum benzoin and white wax; boil two-thirds.

Photo-Mezzotint Engraving.—The following simple method is published :—Upon a polished steel plate spread a thin coating of saturated solution of bichromate of ammonia, 5 drachms; honey, 3 drachms; albumen, 3 drachms; and water, 1½ pints. Let this be dried thoroughly by gentle heat, and expose to light under a transparency. Now remove the plate to a place in which the air is moist. The atmosphere in an ordinary room contains moisture sufficient to act upon the surface of the picture which has been printed in the manner indicated. The preparation given above is slightly deliquescent, and on becoming quite dry by the application of heat attracts so much moisture from the atmosphere as to become more or less tacky. But the exposure to light has the tendency of hardening the film, so that the tackiness produced is in the inverse ratio of the luminous action. A large camel's-hair brush is now charged with a mixture of the two finest kinds of emery powder, and applied with a circular whisking motion all over the surface. As those portions of the plate on which the light did not act are the first to become tacky, the emery powder will first adhere to them, the coarsest particles attaching themselves to those parts of the picture deepest in shadow. The exposure to light ought to be such that every portion of the surface—with the exception of the extreme high lights—becomes in a condition to "take" the powder. If the image be slow in developing under this pulverulent treatment, then the moisture in the atmosphere should be slightly increased. The mere allowing of the picture to stand for five minutes longer frequently answers every purpose ; the moistening of the air by artificial means will answer the same purpose without delay. This film is so susceptible to the influence of moisture that the operator should take great care lest his damp breath impinge on the picture, as the moisture caused by such an application might result in a local predominance of the power which attaches itself in obedience to hygrometic law. Assuming now that the picture has been developed, a polished plate of metal, softer than that upon which the picture is formed, must be laid down upon the other, face to face, and the two passed between a pair of rollers screwed so well together as to insure the setting off on to, or indentation of, the emery powder image into the polished plate of metal. This latter plate is now precisely similar to the one produced by the mezzotint engraver. An impression having been obtained by an ordinary copperplate press, the manipulator having the proof and the plate both before him, applies a small burnisher with a curved point to the various portions of the picture requiring lightening. After having completed this work to the best of his judgment, a second proof is obtained, and, if necessary, a second series of the alterations are made upon the plate until it is finally found that it yields an impression quite equal to the requirements of the subject. It only remains then to hand the plate over to the printer, who will produce the impressions equal in every respect to the first proof.

Etching on Steel.—The metal to be etched is first moderately heated, and whilst hot it is covered with a very thin coating of protective varnish, which adheres all over. This coating, technically called ground, consists of 4 parts asphaltum, 2 parts Burgundy pitch, and 1 part white wax. The engraver then marks out the design, removing the ground from those parts which it is intended shall be etched. The tools used for this process are called etching needles, which are of various thicknesses, so that lines of corresponding breadth can be made easily. A border of wax is then made around the design, forming a trough containing the biting-in acid. This biting-in composition is diluted with nitric acid, according to some authorities; but Knight's Dictionary gives it as equal parts of pyroligneous acid and nitric acid, to which thrice the quantity of water is added. The acid is moved over the surface of the steel by the aid of a feather, so as to act more readily. By this process a certain number of copies may be obtained from the plate, but owing to the fineness of the lines the number is very limited. Imitation etchings are now so well done that at first sight it is difficult to discover the difference.

Polishing Zinc Plates.—The polishing of zinc plates for printing purposes is generally done by a dry process, which, by a German trade paper, is designated as very dangerous to the health of the operator; the fine particles of the zinc and of the emery used for polishing will rise in the air and get into the lungs, and every workman ought to tie before his mouth and nose some cotton cloth, or perhaps, better still, a damp sponge. Another paper gives a wet process of polishing, stating that after the cleaning of the plates by some acid, a wet bit of cloth is dipped into pumice-stone powder, and a few minutes' polishing will suffice to prepare the plates, after which a short finishing polish with fine emery powder is all that is wanted.

To Preserve Zinc Plates.—Plates of zinc such as are used in lithographic and zincographic establishments may be spoiled by a single drop of water being left on them by inadvertence. Mutton fat is an excellent means of preventing oxidation and the influence of humidity. Before using it, the plate must be rubbed perfectly dry with a smooth and clean linen rag; then the fat is lightly rubbed over the surface. When the plate is to be used again, the grease may be easily washed off with spirits of turpentine.

Engraving on Zinc.—Some interesting investigations have recently been made as to the use that can be made of amalgamated zinc. The property of an amalgam of zinc and mercury in repelling fatty ink is already known, drawings made on zinc with mercury being unaltered after an inked roller has been passed over them. If, instead of inking the plate, a very dilute solution of nitric acid (two cubic centimètres of acid in 100 grammes of water) be flowed over it, the amalgam will be dissolved, and the lines consequently be etched in; on the contrary, if hydrochloric acid is used, the amalgam is not attacked, but the zinc is dissolved and the lines are left in relief. Now, proceeding on these principles, if we coat the surface of a zinc plate with some photographic preservative medium, such as bitumen of Judæa or bichromated albumen, the parts where the metal is exposed can then be acted on by mercury, either in a liquid state or in the form of a solution of salt of that metal (like the biniodide), and will be converted into the amalgam, which can be treated so as to produce the image either in relief or in etching, as may be required. Another method of procedure would be to take a copy of a line drawing by the gelatine process, and then dusting the image while still in a humid state will produce biniodide of mercury. This, when dry, would be applied to a plate of zinc, and a copy of the image would be produced on it in amalgam. By then plunging the plate into a bath of nitric or hydrochloric acid, as before, the image will be obtained either in relief or engraving.

Engravers' Plate Marks.—Artists' proofs are without engraved title, but are sometimes signed in pencil by both artist and engraver. Remarque proofs usually have a device in the margin, such as a head, which constitutes remarque; formerly a part of the engraving was left unfinished, as a button or salt cellar, etc. Proofs before letters are still without title, but with artist's and engraver's names printed close to the bottom of the work. Lettered proofs have the title of the work shortly and lightly engraved in a manner capable of erasure before the title is finally placed on the plate in the print state.

To Harden Zinc Plates.

To Harden Zinc Plates.—In order to harden these plates, and thus render them more durable, it has been proposed to nickel them. Zinc, however, takes nickel badly. It is, therefore, recommended to apply first a thin layer of quicksilver, which may be effected by putting the plates into an acidified solution of bichloride of mercury or nitrate of mercury. But the solution must be left to operate only a short time, as otherwise the metal becomes brittle and short. Pure zinc, not amalgamated, is soon coloured yellow and brown in nickel solution ; the deposit may be removed with paper. By employing a weak galvanic current, chemical action preponderates, and a bad deposit is obtained in consequence. If the current is very powerful, zinc is covered more quickly with nickel than a chemical action of the zinc upon the solution can take place, and a good deposit is consequently obtained. In this manner zinc may be nickelled directly. Amalgamated zinc displays a weak action upon nickel solution only after some time, and even a slight amalgamation of the zinc is sufficient to nickel perfectly with a weak current. Considering the extending application of the chemigraphic and photo-chemigraphic process, the above employment of quicksilver will be welcomed by many.

How Etching is Done.

How Etching is Done.—Etching is considered by some as mere pen-drawing, and by others as an inferior kind of engraving. It is, however, an art quite distinct from either, with capabilities and limitations peculiar to itself. Briefly, the process is as follows :—A metal plate, preferably copper, is covered with a coat of blackened varnish or wax. On this surface the artist, with a needle not unlike a common sewing needle, set in a handle, sketches in his composition. The needle usually only removes the varnish, leaving the design in glittering lines upon a black background. The plate is then immersed in an acid bath, and when the lines have been sufficiently bitten it is removed. If variations of tone and a difference of force in the lines are required, as is usually the case, the more delicate portions of the sketch are " stopped out," that is, covered with varnish, so that they shall not be affected by any subsequent exposure in the bath. The plate is again immersed, and the process of stopping out repeated. In the plate by Maxime Lalanne, entitled " Fribourg, Switzerland," for example, the copper was five times subjected to the action of the acid. After three minutes' biting the most delicate lines, indicating the extreme distance, were stopped out, and the plate was exposed for three minutes more. After this the nearer distances were stopped out, and so on with successive portions of the plate, protected from the action of the acid for four, ten, and again ten minutes respectively— making the entire time occupied by the biting process only thirty minutes. It will be seen, even from this cursory explanation of etching, not only that the work is autographic, but that it requires the mastery gained only by thorough artistic training, as well as natural powers of no mean order, to become a master etcher. The hand must be firm and true, the lines must all have meaning, the mind must be clear to grasp essentials, and the whole process must be purely intellectual, as no greater difference in effect can be imagined than that produced by glittering lines on a black surface, on the one hand, and that of delicately graded black lines upon a white background, on the other. A positive process is sometimes used, when the etching appears upon a plate as the black lines upon a white surface, but in this process other difficulties occur, as the lines have to be etched in the order of their depth to ensure the relative amount of biting.

Damaged Woodcuts.—A method of restoring damaged wood-engravings has been recommended. Remove all ink from the damaged part, moisten thoroughly with a solution of potash, and dry the wood again by blowing upon it for several minutes the smoke from a cigar. It is said that an engraving thus treated resumes its former state and may be at once used to print from.

To make Counter-Dies for Stamping.—Cast the counter-die upon the face of the die in type metal, and solder it to the brass backing-piece while in the press in order to get a good register.

Process Engraving.—The outline advertising cuts that appear in the American daily papers are generally done by the Kaolatype process. A sheet of iron is coated with plaster of Paris and the lines scratched down to the base, and the scratch mould is then used as a matrix to make a stereotype plate. The better class of work, however, is done by ink etching.

Etching on Copper.—Line or stippled drawings can readily be produced by the following process:—The copperplate is first coated with bitumen on the turning-table, in the same way as in photo-zincography, and when the bitumen is quite dry. an impression from a lithographic stone on transfer paper is applied to it ; when this paper is removed, we have a copy of the impression in fatty ink on the bitumen surface. This surface is then dusted with fine bronze powder, which adheres to the inked portions and renders them quite opaque. If now the surface be exposed to the light, the bitumen covered by powder will be rendered insoluble ; on the plate being placed in some solution which dissolves the bitumen, the copper will be laid bare in the parts not acted on by the light. These parts can then be etched in by a concentrated solution of iron perchloride, and when the depressions are sufficiently marked, the action of the mordant is arrested. and all the undissolved bitumen is removed. We have in the end an intaglio engraved plate. This process, however, will not answer for the production of drawings with half tones.

New Method of Preparing Blocks for Engravers.—Break an egg and pour a little of the albumen on the block, and then thoroughly rub it in. It does not sink into the wood. This preparation, which is kept on the surface all the time, is as follows :—gelatine, ten grains ; albumen, one ounce ; water, one half-ounce. Rub this well in with the palm of the hand or a brush—the hand preferred ; after this preliminary coating put on a little photina preparation, which is also to be well rubbed in. In five minutes a print from a negative can be taken on this block in the sun ; or by the electric light in about three minutes. The process gives good black tones, like those obtained in platinum prints.

Helioline Engraving.—Hitherto the methods applied by the agency of photography for producing photo etchings on copper or steel plates have only been partially successful, with but little advancement over the earliest efforts made by Niepce, Talbot, and others ; the general character of the work produced being shallow and smooth, lacking sufficient depth, tooth, and ink-holding qualities, so essential for plate printing, and so much so that skilled and prolonged labour is called into play to make them of any commercial value whatever for illustrated publications, even where a small edition is struck off, though aquatint granulations and other devices for producing half-tones—as gauze opaque lines upon glass, etc.—have been resorted to to break up the continuity of the plate surface. To overcome this difficulty, the polished plate is prepared with a thin layer of etching ground or other similar varnish, or a film deposited by the battery, in order that its surface may be rendered impermeable to the action of acid ; then with the ruling machine a course of fine lines is ruled across its face, crossing the same with a second or a third series of lines, if deemed advisable, or with dots made by a roulette. In this condition of the plate, the diamond point having cut through the overlying film, exposed and slightly scratched the surface beneath, if it now be bitten, the etching fluid would at once attack the bright excoriated lines with a clean sharp bite to any desired uniform depth. The small spaces between the intervening lines being protected by the impervious film, necessarily preserves its texture and purity of line. But instead of corroding the plate directly after the ruling, a thin gelatine sheet sensitized with bichromate of potassa is taken, and on this, from a transparent positive, an image is made. This, after being developed by washing in water to remove the unaltered bichromate, is then squeezed in close contact with the plate, or, if preferable, the plate is flowed with a solution of bichromated gelatine, asphaltum, or other sensitive fluid, adaptable to this process. In the latter instance, if the fluids are used, the film is dried and receives the visible or latent image in the usual way known to those who practise those methods. After the exposure, if the bichromated glues have been used, the etching may proceed at once. On the contrary, if a sensitive varnish be employed, a solvent for such varnish is used, according to the character of the agent employed, and the parts in their reciprocal relations being more or less insoluble, the soluble parts in their due proportion acted upon by the menstruum float off, and the resulting picture appears and unfolds itself. In either condition, as described above, the plate is bitten, the acid penetrating through to the ruled lines in proportion to the action of the exposure. First the deeper shadows appear, and so on to the middle tones in their several gradations of texture, with the high lights well marked, till the whole picture is corroded into a finished engraving faithfully represented in line, without the aid of etching tool, stopping-out varnish, or the clever handling of the graver, thus producing a helioline-engraved plate, at once ready for the printing press, capable of striking off thousands of impressions. Should the work in any of its parts appear defective, the process can readily be repeated, by ruling an additional series of lines, and submitting the plate to a second exposure, or by rebiting.

Doorplate Engraving.—The usual way is to draw the forms of the letters upon the plate with a steel point or a pencil, and dig out the letters with a graver. The filling matter is black or red sealing wax. A way of etching the letters with acid has lately come into practice. With a complicated design some very pretty work is done in this way. The next is machine engraving, one kind being done by a routing machine carrying an automatic tracer traversing the pattern. Of these engraving machines there are several in the market under various patents, some as mere tracers, others as liners, while some claim universal work.

Writing on Metals.—Take half a pound of nitric acid and one ounce muriatic acid. Mix and shake well together, and then it is ready for use. Cover the plate you wish to mark with melted beeswax; when cold write your inscription plainly in the wax clear to the metal with a sharp instrument. Then apply the mixed acids with a feather, carefully filling each letter. Let it remain from one to ten hours, according to the appearance desired, throw on water, which stops the process, and remove the wax.

To Print an Etching.—The process is as follows :—
The plate is first warmed by being laid on a sheet of iron,
under which small gas jets are constantly kept burning ; then
the ink is spread on to the surface and into the lines with a
dabber, the superfluous ink being wiped off with a coarse
muslin rag, care being taken not to wipe the ink out of the
lines while removing it from the surface. Simple as this
appears when done by a practised hand, it really requires
considerable skill. The palm of the hand is then rubbed over
with a little whitening, and a final polish is given to the plate
with it. The plate is now put on the travelling bed of the
press, and on it is laid the paper, which should have been
previously damped ; over all are laid several thicknesses of
flannel. On the handle of the press being turned the cylinders
revolve, and the travelling board passes between them carrying the plate with it. By the pressure thus obtained the
paper is drawn into the lines on the plate, the process being
facilitated by the elasticity of the flannel. Care must be
taken not to tear the paper in removing it from the plate.

Photography for Wood Engraving.—Hitherto the
attempts which have been made to directly use photography
for the purposes of wood engraving have not met with entire
satisfaction. The difficulty lies in the fact that the film upon
the surface of the wood is apt to be torn by the graver employed in cutting. By a new method the film is scarcely per-
ceptible, notwithstanding a collodion image is impressed upon
the wooden block. For the fixing of this image alcohol is
employed instead of water. A collodion diapositive is prepared,
covered with glass, and laid in a bath of alcohol, several blocks
being laid one upon the other. Under these conditions the wood
does not swell up as would be the case where water is employed,
and the collodion adheres very closely to the wood.

Furniture needs Cleaning as much as other woodwork. It may be washed with warm soapsuds quickly, wiped
dry, and then rubbed with an oily cloth. To polish it rub
with rotten-stone and sweet-oil. Clean off the oil, and polish
with chamois-skin.

Etching on Glass.—A liquid which can be used with an ordinary pen, can be made of equal parts of the double hydrogen ammonium fluoride and dried precipitated barium sulphate, ground together in a porcelain mortar. The mixture is then treated, in a platinum, lead, or gutta percha dish, with fuming hydrochloric acid, until the latter ceases to react.

Etching Metal Surfaces.—The following method of etching metallic surfaces, by which it appears possible to produce highly decorative effects, has recently been published. The article to be treated is electroplated with gold, silver, nickel, or other metal, and on this the design which it is desired to produce is traced with some suitable acid-resisting substance. It is then immersed in an acid bath, by the action of which those portions of the surface which are left unprotected are deprived of their electroplated coating, and the naked metal beneath is given a frosted or dead appearance. The article is then well rinsed to remove all traces of the acid employed, and the acid-resisting varnish is removed by the use of alcohol, oil, or other proper solvent. The result is a frosted or dead-lustre surface of the original metal, upon which the design in the electroplated metal stands up in relief. If, for example, the article be one of copper and the plating silver, the design will be in silver upon a dead copper ground. It is manifest that the operation may be reversed— that is, the design to be reproduced, instead of being protected, as in the foregoing procedure, may be left unprotected, and the remainder of the electroplated surface covered. In this case, after going through the above-described operations, the design would appear to be in dead copper on a silver ground.

Oilcloth may be improved in appearance by rubbing it with a mixture of a half-ounce of beeswax in a saucerful of turpentine. Set this in a warm place until they can be thoroughly mixed. Apply with a flannel cloth, and then rub with a dry flannel.

BOOKBINDING.

WOOD for Binding.—A novelty has been introduced in bookbinding—the substitution of walnut, maple, mahogany, sycamore, and cedar for the covers of reading-cases, music-books, and large volumes, in place of the usual leathers, muslins, or papers. The examples submitted in Gothic binding of this kind contrast very favourably, both in cost and appearance, with the ordinary leather, and, so far as we can judge without absolute trial, seem likely to be well received for table-books, folios, and albums. The several woods, which take a high degree of polish, are prepared to resist stains and damp, and the cost, we are informed, does not exceed that of leather binding.

The Covers of Books.—The vexing question of binding will come after all the other questions have been settled. It shall not be panelled russia, certainly, for russia leather does not last, but it may be crushed levant, morocco, or vellum over the original paper wrappers and the original cloth covers detached from their boards, or the cloth-covered books may remain untouched; and your binding may be half binding with corners, or half binding without corners, or full binding; and in Jansen (short for Jansenist) style, which is without an ornament, or *à la fanfare*, which is a pretty style of ornamentation, and in one or varied colours; and it is not heresy in bibliolatry to give an English book to a French bookbinder, or the reverse, and an American bookbinder loves his art enough not to let his Chauvinism make ungraceful work for "American Notes." The only recommendation which it would not be pedagogic to make to a collector may be that a book in half binding should have uncut edges, and a book in full binding must have gilt edges, even if the binder finds it necessary to cut them a little, for full binding is the evening dress of books, and a book in a full binding, with or without gilt top, and white edges on the other sides, as it often appears recently, looks like a gentleman in a clawhammer of broadcloth and blue jean trousers. The paper covers (or wrappers, as the Dickens men say) of a book are kept because they are often illustrated, were part of the book's physiognomy, mayhap made its fortune, and contain useful information for bibliographers. The French led the fashion with their first collections of books written by romanticists, the covers of which had ghostly vignettes by Deveria and the Johannots, and magnificent promises of great works by Hugo, Gautier, Balzac, Lacroix, and the rest, a catalogue of which would make an interesting volume on " Books Announced, but Never Published."

" Pellisfort " Binding.—Something new in bookbinding has been introduced by Messrs. Stoneham and Co., of Farringdon Road, who have been very successful in bringing out quite a new feature in this line of business. It is called the "Pellisfort" binding. Instead of the ordinary boards metal plates are used, which for many reasons are found to be very much better. The covers are flexible and yet strong, and will not curl up even if the book be read close to the fire, and they also resist damp. A better finish, it seems, is given to books bound in this style. The leather lays flatter and the gold looks brighter. Many people prefer limp binding, and this " Pellisfort " style of binding not only gives limpness, but strength and elegance combined. Another excellent feature we noticed in the bound books shown us was a place made in the inside of the front cover for a photograph, which for presentation volumes is a capital idea. There is no doubt that this new binding will take well with the trade and the public. The above firm are also making " Pellisfort " leather photo frames on the same principle, metal plates being used instead of boards. They are covered in morocco and made to stand on a shelf, or can be hung up; they are finished in a very superior style.

Embossed Leather.—Hundreds of men, women, and even boys, in New York, are engaged in the business of collecting old boots and shoes, which they take to the wall-paper factories, where they receive from five to fifteen cents per pair. Calfskin boots bring the best price, while cowhide ones are not taken at any figure. These boots and shoes are first soaked in several waters to get the dirt off, and then the nails and threads are removed and the leather is ground up into fine pulp. Then it is pressed upon a ground of heavy paper, to be used in the manufacture of " embossed leather." Fashionable people think they are going away back to mediæval times when they have the walls of their libraries and dining-rooms covered with this, and remain in blissful ignorance that the boots and shoes which their neighbours have thrown away now adorn their walls and hang on the screens which protect their eyes from the fire. Carriage-top makers and bookbinders also use the material made from old boots and shoes, the former for leather tops for carriages and the latter leather bindings for the cheaper grade of books. The new styles of leather frames, with leather mats in them, are entirely made of the cast-off feet coverings.

Colours of Books in the British Museum.—These are all bound on a principle; historical works being in red, theological in blue, poetical in yellow, natural history in green. Besides this, each part of a volume is stamped with a mark by which it can be distinguished as their property, and of different colours; thus, red indicates that a book was purchased, blue that it came by copyright, and yellow that it was presented.

Mildew in Books.—Readers whose books have been, or are likely to be, attacked by mildew, may preserve them to some extent by placing a saucer of quicklime near, in the bookcase, or shelf, or where convenient. The lime absorbs the excess of moisture, and must be renewed, as it becomes slaked and loses its strength. It is equally good for putting in linen chests, iron safes, or wherever there is likely to be any mustiness owing to the exclusion of fresh air.

Sizes of Books (Uncut Edges).—Books with uncut or merely trimmed edges should measure in inches:

	Octavo.	Quarto.
Pott	$6\frac{1}{4} \times 4$	$7\frac{3}{4} \times 6\frac{1}{4}$
Foolscap	$6\frac{3}{4} \times 4\frac{1}{4}$	$8\frac{1}{2} \times 6\frac{3}{4}$
Crown	$7\frac{1}{2} \times 5$	$10 \times 7\frac{1}{2}$
Post	8×5	10×8
Demy	$8\frac{3}{4} \times 5\frac{1}{2}$	$11\frac{1}{4} \times 8\frac{3}{4}$
Medium	$9\frac{1}{2} \times 6$	$12 \times 9\frac{1}{2}$
Royal	$10 \times 6\frac{1}{4}$	$12\frac{1}{2} \times 10$
Super Royal	$10\frac{1}{4} \times 6\frac{3}{4}$	$13\frac{3}{4} \times 10\frac{1}{4}$
Imperial	$11 \times 7\frac{1}{2}$	15×11

Other sizes are a matter of further subdivision. These dimensions are not for books with cut edges, but it is safe to allow a quarter of an inch less in height, and not quite so much in width, if the edges have been cut down, always assuming these edges have not been cut more than once; otherwise, if the book has been rebound more than on one occasion, no reliance can be placed on this rule.

Waterproof Binding.—A new waterproof material, designed as a substitute for bookbinders' cloth or leather, is made by treating fibres of linen, cotton, jute, wool, wood, cellulose, etc., as well as the waste from the treatment of such fibres or cellulose, with a solution of albumen, which solution may be applied superficially or throughout the mass. To this solution is added glycerine, in order to make or render the coagulum formed ductile, and a solution of metal salt for the purpose of increasing the capability of coagulation, and for preserving it against the action of water; and in order to impart to the material the strength of leather or stiff paper of strong quality, basic borate of soda is added. The solution of albumen thus prepared is brought into coagulation, and the whole is pressed in an appropriate manner between heated rollers or cylinders, which may have smooth or engraved surfaces as desired. Instead of using fibres as described, woven material or paper may be employed, using the same process, roughening the surfaces of the material to facilitate the adhesion of the albumen and keep the coagulum on the surface.

s

The Manufacture of Gold Leaf.—The following is the method usually adopted by manufacturers of gold leaf. The extreme malleability of gold has made it a prominent metal in the useful as well as the fine arts. It has been calculated that from two to four million dollars' worth of this precious metal is annually used on ornaments, in gilding, in lettering, edging of books, in sign, ornamental painting, and in dentistry. Of course, the greater proportion of this is for the first-named purpose, although there appears to be a great amount used in the other industries... A comparatively small amount, however, is necessary to cover a great deal of space. A cubic inch can be hammered so as to cover a space thirty-five feet wide and one hundred feet long, and twenty twenty-dollar gold pieces can be drawn into a wire that would reach around the globe.

A New Cover for a Writing-Book.—This is made with a pivoted rod which falls into the hollow of the book, and to which is attached an adjustable swinging-arm upon which the copy slip may be placed, and which arm also serves to keep down the leaves.

Leatheroid.—A new article made of paper. It consists of a number of thicknesses of cotton paper wound one upon another over a cylinder. It possesses remarkable qualities of strength and adhesion, derived from a chemical bath through which the paper is drawn on its way to the cylinder. Leatheroid, for the purposes it now serves, consists of about twenty thicknesses of paper ; it is shaped upon or around moulds, while wet, into the form it is to represent, and will hold that form perpetually when dry. When dried, it is as difficult to cut with a knife as raw hide. A company has been formed for the manufacture of this article in the United States, and for its manufacture into seamless cans, boxes, etc., to take the place of tin cans and wooden boxes. Cans made from it are about one-fourth the weight of tin cans of equal size. Cans made from leatheroid have the elasticity of thin steel, and no amount of kicking or handling will break them,

To Preserve Bindings from Mildew.

—The binding of books may be preserved from mildew by brushing them over with spirits of wine. A few drops of any essential oil will secure bindings from the consuming effects of mould and damp. The leather which is perfumed with the tar of the birch-tree never moulds or sustains injury from damp or insects. The Romans used oil of cedar to preserve their manuscripts. Russia leather placed in the window will, it is said, destroy flies.

Persian and Turkey Morocco.

—According to "The Library Journal," Persian goat or morocco is the skin of a kind of wild goat raised in East India, and tanned in a species of bark native in its own country, and then shipped to London, from which place it is sent to all parts of the globe. Turkey morocco is a goat-skin raised in Switzerland and sent to Summac, Germany, for tanning, and is a finer grade of goods. Bock leather is a sheep-skin, also raised and tanned in East India.

Pig-Skins for Bookbinding.

—At one time, we believe, before morocco came into general use, pig-skins were largely employed for bookbinding. One reason why they fell into disuse, probably, was their cost, but this, at least, is no longer an impediment. Pig-skins are tanned by an old-fashioned and somewhat tedious process, but it has the advantage of turning them into a leather of extraordinary toughness and durability, which is all but impervious to atmospheric influences. We lately saw some specimens of pig-skins, or, as they are named by the tanners, hog-skins, specially manufactured by Messrs. John Muir and Sons, of Leith, for the use of bookbinders. They are in a variety of shades, and are worked up into a very beautiful grain, something like a large grain morocco in appearance. Their cost is a little below that of calf, so that they fall quite within the limits of expense as materials for bookbinding. Appearances are greatly in their favour, and whether for library bindings, where strength and durability are the prime considerations, or for fine bindings, these hog-skins seem equally well adapted. Hog-skins would be very serviceable for account-book bindings, particularly for hot climates. In India and elsewhere hog-skins have long been preferred to other leathers for certain purposes, on account of their lasting qualities, and similar considerations may be expected to operate in their favour for bookbinding. One curiosity of hog-skin is its resemblance to human skin; after being tanned the one can scarcely be distinguished from the other.

Margins of a Book.—The four sides of a printed page in a book are called head, tail, fore-edge, and back.

Edges of Books.—There is not always a clear understanding as to the terms used in connection with the treatment of edges. "Uncut" does not necessarily mean that the edges have not been opened with a knife, but simply that the book has not been cut down by machine, a method which sometimes sadly mars the appearance of a book. The expression "unopened" is perhaps a stricter term to be applied when absolutely untouched. "Trimmed edges" means that the heads have been left untouched, and the fore-edge and tail merely trimmed sufficiently to make them tidy. "Cut edges" means that a portion has been cut from the three sides of the book.

Some Bibliographical Names of Books :

Folio	2°.
Quarto	4to.
Octavo	8vo.
Duodecimo	12mo.
Sextodecimo	16mo.
Octodecimo	18mo.
Vigesimo-quarto	24mo.
Trigesimo-secundo	32mo.

A Remedy for Soiled Books.—Dirt may be taken off book-leaves, without injuring the printing, with a solution of oxalic acid, citric and tartaric acid. These acids do not attack printing-ink, but will remove marginal notes in writing-inks, etc.

Imitation Bronze Plates for Book Covers are produced by embossing a metal-covered foundation plate, and bronzing and varnishing the surface so as to give it the appearance of old bronze or other metal. A foundation plate of paper-pasteboard or other suitable material, faced with silk, muslin, or cloth, is covered with a thin layer of metal, formed of metal leaf, metal-covered paper, foil, tinsel, or bronze powder. The plate is then embossed by means of an embossing plate into which the required design has been engraved or otherwise produced. The surface of the foundation plate is next sized and coated with pulverized graphite, bronze powder, or other metallic powder, over which a coat of varnish is placed to protect the bronze coating. The plates thus produced have the appearance of real bronze, present a handsome appearance, and can be used for a variety of purposes. By using bronze powders of different colours on the same plate a variegated effect may be imparted and the artistic appearance of the plate considerably enhanced.

Book Labels.—Amateur bookbinders would produce tasteful labels for the backs of books if they only knew how to impress the letters on the thin leather in such a way as to prevent the gold from rubbing off. To print in the ordinary manner and bronze over is useless. The best plan is this: Thoroughly beat the white of an egg, rub it thin over the place to be lettered, put on the gold leaf, and with type, sufficiently heated to coagulate the albumen, press on the leaf. Remove the surplus leaf with a tuft of cotton.

Red Edges.—To obtain a bright and lasting red edge take the best vermilion and add a pinch of carmine; mix with glaire, slightly diluted. Take the book and bend over the edge so as to allow the colour to slightly permeate it; then apply the colour with a bit of fine Turkey sponge, bend over the edge in the opposite direction, and colour again. When the three edges have been done in this manner, allow them to dry. Screw the book tightly up in the cutting-press, and after wiping the edge with a waxed rag, burnish with a flat agate.

A Preventive against Dampness for Books.—

Quicklime is the best thing to save books from the ill effects of damp. A small vessel full of lime placed near a bookcase is better than a blazing fire for this purpose. The lime, which absorbs every particle of moisture in the atmosphere, must be changed every two or three days.

A Strong Binding.—

The following method of binding books liable to rough usage has been patented by a resident of Switzerland. After the sheets have been folded, collated, and pressed, they are glued on the back. The blank paper, glued inside on both parts of the binding boards, receives linen folds. After these blank papers have been glued or pasted, the book is trimmed on the three sides and a board-lap glued along the back. The book is then glued into its binding. The holes for the rivets are punched by a machine, and little metal plates placed along both sides of the back of the book are riveted with wire having little heads. The sewing of single sheets, which involves great loss of time, is done away with in this binding.

Morocco Leather.—

Morocco leather is made from goat-skins, tanned in sumach, dyed in the ordinary way, having been previously immersed in a solution of sulphuric acid. The grain is stamped upon it either by hand or by machinery, similar to that employed for the purpose of dicing or graining. Very fine small skins for gloves are often prepared by immersion in a solution of alum and salt, instead of tannin, flour and the yolk of eggs being afterwards applied to soften and whiten. Buff leather, not now quite so much in request as in former days, was at first made from the skin of an animal called the buffe, or urus, which was then common in Western Europe. When new the leather was always a tawny yellow, and the skins gave the name to the colour. Cordovan leather was first made at Cordova in Spain, from hides dressed to be used with the grain side outward. It was from this leather that the title " cordwainer " came. Russia leather is tanned in an infusion of willow or birch bark, and derives its peculiar and long-enduring odour from the birch oil with which it is dressed. Levant leather is first "struck out" in warm water on a mahogany table, "blacked" with logwood and iron liquor, then polished by revolving rollers, and "grained up" by the workman with a "corking board" on a table. The grain is set into the leather in a hot stove, and after this it is oiled with cod oil. In finishing japanned leather the japanned mixture is worked by the hand alone. This mixture consists simply of linseed oil and Prussian blue, the last coat being of linseed oil and lampblack, put evenly over the surface as it lies spread out on a table. No machine has as yet been made to supersede the hand in this part of the work. In the blacking of skins a mixture of ox blood and acetate of iron is now very often used.

Preservation of Bindings.—For this purpose the use of vaseline has been suggested, it having the advantage of being a mineral substance, and therefore much less liable to decompose than anything belonging to the animal or vegetable kingdom. It answers better, however, with leather and with cloth than with the marbled sides or edges of books, though even these have been found not to be in any way damaged by the treatment. It might be thought that an unpleasant greasiness would be produced, but this is not so—at least not for more than a few hours. The bindings seem to drink up the vaseline as if they knew it would do them good. Neither does the smell of vaseline persist for long. At the same time it is well to be cautious, and anyone who is disposed to make trial of the plan here recommended would in the first instance do well to confine his attentions to old bindings.

A Cheap Binding.—Pack the papers smoothly, hold firmly, and drive a thin chisel through the pile about half an inch from the back. Push a strong tape through and leave out about two inches; put three or four tapes through at even intervals. Cut common thick paper boards large enough to project a little everywhere, except that one edge must come in front of the tapes. Draw the tapes tightly, and glue down to the boards outside. Skive a piece of leather—common sheep-skin will answer—wide enough to cover the back and come on the boards an inch or two, and long enough to project a couple of inches at the end. Paste the leather well; put it on the back; fold the ends in so as to come over the boards on each side. Paste any fancy or plain paper over the sides; and, lastly, paste the blank leaf down to the cover inside, and you have a presentable book, and very durable. Trimming the edges can be easily done by clamping between boards and cutting the edges with a thin, sharp knife by a straight edge. Of course this is done before the boards are put on, after the tapes are in. This makes a flat-edged book, but for a thin book answers very well.

To Gild Edges of Cards.—Put your cards together so that the edges are perfectly even. Then place in a press with the edges uppermost. Coat the edges with a mixture of red chalk and water. The gold leaf—preferably of thin quality—is blown out from the books and spread on a leather cushion, where it is cut to the proper size with a smooth-edged knife. A camel's-hair pencil is dipped into white of egg mixed with water, and with this the partially dry edge is moistened; the gold is then taken up on a tip-brush and applied to the moistened edge, to which it instantly adheres. When all the four edges have been gilt in this way, and allowed to remain a very few minutes, take a burnisher formed of a smooth stone (usually bloodstone) and rub the gold quickly till it receives a high degree of polish. For silver edges take a brush, dip it in a saturated solution of gallic acid, and wash the edges; then brush with a solution of 20 parts nitrate of silver to 1,000 parts distilled water. Keep on alternating these solutions until the edges assume a brilliant tint. Then wash with distilled water and dry by free air and heat.

A Process for Marbling Book-edges.—Marbling the edges of books can be done, and as a matter of fact frequently is done, by practical bookbinders simply by transferring from unglazed marbled papers. The operation is conducted by first screwing the book to be marbled up tight in a press and damping the edges with spirits of salt. The marbled paper selected is then laid on and another piece of moderately stout paper on the top, and is then carefully beaten down with a broad smooth-faced hammer, taking care that the paper does not slip while doing so. After going over the edges with light but decided blows several times the pattern will be found perfectly transferred and may then be burnished. To produce good results all that is required is a little care and practice.

Remedy for Wood-worms.

Remedy for Wood-worms.—Benzine will destroy wood-worms in books and woodwork. The insects, as well as their larvæ and eggs, will soon die off if a saucer of benzine is placed in the bookcase and the doors kept closed. Furniture and carvings infested with wood-worms are similarly placed in a room with a dish of benzine, and kept closed up for a sufficient time.

Gilding Upon a Marble Edge.

Gilding Upon a Marble Edge.—The book, being scraped the same as in gilding, is marbled in a desired style and put into the gilding-press in the same manner as an ordinary book for gilding ; it is then burnished, the size being lightly applied immediately, care being taken so as not to destroy or unsettle the marble, and then is finished off in the ordinary way, after which, when properly done, it makes as fine an edge as one would wish to see, the marble showing through the gold.

Glass for Paring On.

Glass for Paring On.—Use a piece of glass instead of marble ; you will find it much to be preferred. A piece of heavy plate-glass is the best, as it will not break so easily.

Colour for Red Edges.

Colour for Red Edges.—A very pretty red edge can be made of Chinese vermilion and eosine mixed. First dissolve the vermilion in the usual way and then add the eosine powder—about half as much eosine as vermilion—and when burnished the edge will have a gilt or metallic cast.

Imitation Leather Surfaces.—By means of electricity the most attractive leather surfaces are now completely imitated. The leather which it is desired to imitate is first well cleaned and coated with graphite, as in electroplating a small article. It is then placed in a copper-bath, the tank of which is large enough to easily receive a skin of any size. A dynamo-electric machine generating a powerful current furnishes the electricity. The copper is deposited upon the coated surface of the hide to a thickness of from one-sixteenth to one-eighth of an inch. The plate thus formed reproduces, but reversed, every mark and minute vein of the leather, so that a print taken from it is an exact copy of the original in every detail.

Black Under Glaze.—Red lead, 3 parts; $1\frac{1}{4}$ parts antimony; $\frac{3}{4}$ part manganese. After these have been calcined, add the following, and calcine again:—3 parts blue calx ; $\frac{1}{4}$ part oxide of tin.

Pearl Patterns on Cloth.—Flexible mother-of-pearl patterns are produced on cloth stuffs, according to a recent German patent, as follows:—On a soft elastic base is placed thin caoutchouc as large as the pattern, and upon this a thin plate of copper, with the pattern cut through. Over the copper is placed the cloth on which the mother-of-pearl pattern is to be produced. A heater is now passed over the whole, with the result of melting the thin caoutchouc, and causing it to be pressed up against the cloth, in form of the pattern.

Cement for Leather.—Not every bookbinder may be aware that gutta percha dissolved in carbon disulphide until it is of the consistency of treacle forms a very good cement for splicing leather. The parts to be joined must first be thinned down, then a small quantity of the cement is poured on each end, and spread so as to thoroughly fill all the pores of the leather. The parts are next warmed over a fire for a few minutes, joined quickly, and hammered well together.

Imitation Leather used for binding cheap account books and other purposes, in place of leather, is manufactured thus:—A piece of cotton texture is passed between two cylinders, the upper one of which permits a mixture consisting of oil, resin, lampblack, and other matters to flow upon the slowly moving canvas. From the cylinders the fabric is wound up upon a drum made of wooden sticks so arranged that the successive layers are kept apart from one another. When the whole piece has been wound up on the drum, the latter is placed, with the oiled cloth on it, in a drying chamber. After drying, the cloth is smoothed by means of a pumice-stone, and passed a second time through the cylinders, receiving another coating of varnish. It is then dried, and these alternate operations repeated at least five times, in order to make the coating sufficiently thick. The final process is pressing the cloth in such a manner as to give it the appearance of natural leather.

To Restore the Gloss of Fine Bindings.—These may be restored by using a preparation of Canada balsam and clear white resin, of each six ounces, dissolved in one quart of oil of turpentine.

Gilding on Water-Coloured Edges.—This is an edge which not only pleases the eye but is hard to execute. After the edge is well scraped and burnished, the leaves on the fore-edge are evenly bent in an oblique manner, in which position boards are affixed on each side until the design is painted, according to fancy. When dry, the boards are untied and the leaves take their proper position. The book is then placed in the gilding-press, and the size and gold are laid on, and when dry burnished. The design will not be apparent when the book is closed, because the gold covers it; but when the leaves are drawn out it shows up readily, the gilding disappearing, and a very unique effect is produced. The time and labour required make this operation expensive, and it is consequently very seldom done. However, the taste and wishes of some render it necessary that the gilder should know how to operate.

Protection against Cockroaches.—These insects are known to be great destroyers of books in the ravages they make upon the bindings. Roaches will not touch books which are varnished with a mixture of one part copal varnish and two parts oil of turpentine. With a large brush paint this over the cloth binding, and let the book stand to dry. It cannot be applied to the edges, unfortunately, but it is something to know that it will save the other parts of the book.

Gilding Leather.—Damp the skin with a sponge and water, and strain it tight with tacks on a board sufficiently large. When dry, size it with clear double size; then beat the whites of eggs with a wisp to a foam, and let them stand to settle; then take books of leaf silver and blow out the leaves of the silver on a gilder's cushion; pass over the leather carefully with the egg size, and with a tip-brush lay on the silver, closing any blister that may be left with a bunch of cotton. When dry, varnish over the silvered surface with yellow lacquer until it has assumed a fine gold colour. The skin being thus gilded may be cut into suitable strips or patterns. It should be carefully observed to have the skin well dry before sizing it. Bookbinders gild the edges, etc., of leather in a different way. They first go over the part intended to be gilded with a sponge dipped in the glaire of eggs (the whites beaten up to a froth and left to settle); then, being provided with a brass roller on the edge of which the pattern is engraved, and fixed as a wheel or roll in a handle, they place it before the fire till heated, so that, by applying a wetted finger, it will just hiss. While the roller is heating they rub the part where the pattern is intended to come with an oiled rag or clean tallow, and lay strips of gold-leaf on it, pressing it down with cotton; then with a steady hand they run the roller along the edge of the leather, and wipe the superfluous gold off with an oiled rag, and the gold adheres in those parts where the impression of the roller has been made, while the rest will rub off with the oiled rag.

To Restore Morocco.—The lustre of morocco leather is restored by varnishing with white of egg.

Restoring Leather Bindings.—Mildew, shown in the form of roundish or irregular brown spots, cannot be cured, but may be effectually checked by thoroughly drying the volume and afterwards keeping it in a dry place. Leather bindings of old books will frequently be found to be dilapidated. If broken, rubbed, or decayed, fill up the crevices with paste, then take the yolk of an egg, well beaten up, and apply to the leather with a sponge, having first cleansed it with a dry cloth. A hot iron passed over it gives a polished surface. Stains of any kind may be removed either from the leaves or the cover of a book. For common writing ink use a mixture of spirits of salts and water in the proportion of one to six. A solution of chloride of lime is also good. In both cases the part should be subsequently well washed with clean water. Grease or wax spots are removed by holding a hot iron close to the place affected, or by washing with ether or benzine. By the latter process it is advisable also to use the hot iron. The remedy against oil stains is sulphuric ether. Roll up the leaf to be operated on, insert it in a flat-mouthed bottle half full of the ether, and shake it quietly up and down for a few minutes. On removal the stains will have disappeared; the ether rapidly evaporates from the paper, and rinsing in a little clear water finishes the job.

To Preserve Records and Books against the Attacks of Insects.—It has been proposed:—1. To abolish the use of any wood in the binding processes. 2. To recommend the bookbinder to use glue mixed with alum in place of paste. 3. To brush all wormeaten wood in the repositories of books with oil or lac-varnish. 4. To preserve books bound in calf, it is recommended to brush them over with thin lac-varnish. 5. No book to lie flat. 6. Papers, letters, documents, etc., may be preserved in drawers without any danger, provided the wafers are cut out, and that no paste, etc., is between them. 7. The bookbinder is not to use any woollen cloth, and to wax the thread. 8. To air and dust the books often. 9. To use laths, separated one from the other one inch, in place of shelves. 10. To brush over the insides of bookcases and laths with lac-varnish.

Black Edges.—Books of devotion are usually bound in black leather, and the edge of course is blacked to match the cover. To give a book a neat black edge observe the following process :—Put the book in the press as for gilding, and sponge with black ink; then take ivory-black, lamp-black or antimony, mix well with a little paste, and rub on the edge with the ball of your hand till it is perfectly black and a good polish is produced. Then burnish the same as any other coloured edge. The edges require the book to be scraped in the same manner as for gilding. To lay the colour on evenly and give it a high burnish requires more labour than gilding and is quite as expensive.

Black for Binders.—Brunswick black, thinned down with turpentine until it has attained the right tone and colour, with a little varnish added—about one-twentieth of the bulk of the black and turps. There is no difficulty in getting the mixture to dry hard.

Imitation Leather.—An economical substitute for rough calf or other skins used in bookbinding has lately been invented. Vellum cloth or some other suitable fabric is coated with an adhesive substance, such as is used in making flock-paper, and before this substance becomes dry, flock is dusted upon it. The resulting fabric resembles rough calf or other leather ; the effect can be varied according to the particular dye previously applied to the flock.

A Few Hints on the Care of Books.—Amongst others, the following suggestions to book-lovers appeared in "Notes and Queries," of an old date :—

Never cut up a book with your finger, nor divide a printed sheet if it be ill-folded, as one page will rob the other of its margin.

Never bind a book wet from the press, as it cannot certainly be made solid without risking the transfer of ink from one page to another.

Never compress a book of plates in binding, as it injures the texture of the impressions.

Never destroy an antique binding if in moderate condition ; if necessary repair it carefully. Do not put a new book in an antique jacket, or *vice versâ*.

Never allow a book to be " finished " without the date at the tail of the back ; it saves subsequent trouble, and the book from much needless handling.

Never have registers or strings in your books of reference, they are apt to tear the leaves ; paper slips are the best, if not to numerous.

Do not allow your books to get damp, as they soon mildew.

Do not allow books to be very long in a too warm place. Gas affects them very much, Russia in particular ; Morocco stands heat best.

Rough-edged books suffer most from dust. Gilt edges are the best ; at least, gild the top edges.

Do not, in reading, turn down the corners of the leaves ; do not wet your finger to turn a leaf, but pass the fore-finger of the right hand down the page to turn over.

Do not allow foreign substances, crumbs, snuff, cards, botanical specimens, to intrude between the leaves.

Do not stand a book long on the fore-edge, or the beautiful level on the front may sink in.

Never wrench a book open if the back is stiff, or the edges will resemble steps of stairs for ever after ; open gently a few pages at a time.

Never lift books by the boards, but entire.

Never pull a book from the shelf by the head-band ; do not toast them over the fire, or sit on them, for " Books are kind friends. we benefit by their advice, and they reveal no confidences."

MISCELLANEOUS.

TO Transfer Printed Matter to Glass.—Flow the glass plate with a good quantity of photographers' negative varnish, which should be thinned down in the usual way. When this has been partially dried so that it will not run into the paper, lay the engraving or showbill face downwards upon the prepared surface, and subject it to slight and uniformly apportioned pressure for twenty-four hours. Then moisten the back of the paper, and by means of a soft rubber rub off the softened paper. If this is done with care, the ink lines will remain attached to the varnished glass surface. As the thin varnish is quite transparent, this is equivalent to transferring the engraving to the glass surface. The transfer is frequently improved in appearance by giving the plate (and transfer) a second coat of varnish.

A cheap Gilding Bronze.—The following mixture has been recommended for cheaply gilding bronzes, etc. :—2½ lbs. cyanide of potash, 5 oz. of carbonate of potash, and 2 oz. cyanate of potass, the whole diluted in 5 pints of water, containing in solution one-fourth ounce chloride of gold. The mixture must be used at boiling heat, and after it has been applied, the gilt surface must be varnished over.

To Separate the Leaves of Charred Books.— Cut off the back so as to render the leaves absolutely independent from one another, then soak them, and dry them rapidly by a current of hot air. The leaves will then separate, but must, of course, be handled with extreme care.

Japanners' Gold Size.—In gold printing japanners' gold size may, it is suggested, profitably take the place of the size ordinarily used. Powder finely of asphaltum, litharge, or red lead, each 1 oz.; stir them into a pint of linseed oil, and simmer the mixture over a gentle fire, or on a sand-bath, till solution has taken place, scum ceases to rise, and the fluid thickens on cooling. If too thick when cool, thin with turpentine.

To Keep Brass from Discoloration.—It is difficult to make yellow brass articles long retain their colour, but a thin varnish of shellac or a coating of collodion will preserve the colour without giving the appearance of being varnished.

Colours for Bronze.—In making colours for bronze, manufacturers have hitherto employed a concentrated solution of gum arabic for grinding the bronze, reducing it to powder by pounding. But someone has recently found a better and cheaper material by substituting for the gum arabic a liquid solution of 5 parts of dextrine, and 1 part alum, the bronze being washed and polished as usual.

To Clean Bronze.—Weak soapsuds or aqua ammonia will clean bronze statuary or bronze ornaments in the fine lines of which dust has collected.

To Nickel Plate.—A simple method for nickel plating is to add to a solution of nitrate of nickel a strong solution of potassium cyanide, until the precipitate of the first is dissolved. Nickel anodes must be used. Another bath for nickel deposits is obtained by dissolving equal weights of ammonia and nitrate of nickel, and then mixing the solution with twenty-five times its volume of bisulphite of soda.

To Remove Dust from Polished Surfaces.—Take cyanide of potassium, 15 grammes; soap, 15 grammes; chalk blanc de Meudon, 30 grammes; water sufficient to make a thick paste.

T

To Keep Gilt Goods Bright and Clean.—Stationers

always have some loss in gilt goods, and it is no easy matter to keep these gilt goods bright and clean. Gilt articles, if of metal, may be cleansed by rubbing them gently with a sponge or soft brush moistened with a solution of half an ounce of potash, or an ounce of soda, or perhaps best, an ounce of borax in a pint of water ; then rinsing them in clean water and drying with a soft linen rag. Their lustre may be improved, in certain cases, by gently heating them and then applying gentle friction with a soft rag. A very dilute solution of cyanide of potassium, applied in the same manner, will answer the same purpose, washing in water and finally drying by gentle friction with a linen rag ; but as this substance is very poisonous, it is not to be recommended for household uses. Gilt frames of pictures, mirrors, etc., should never be touched with anything but clean water, gently applied with a soft sponge or brush.

To Clean Brass.—Make a mixture of one part common

nitric acid and one half part sulphuric acid in a stone jar ; then place ready a pail of fresh water and a box of sawdust. Dip the articles to be cleaned in the acid, then remove them into the water, after which rub them with sawdust. This immediately changes them to a brilliant colour. If the brass is greasy it must be first dipped in a strong solution of potash and soda in warm water. This cuts the grease so that the acid has the power to act.

To Clean Nickel.—Nickel silver mounts and ornaments

can be kept bright by rubbing with a woollen cloth saturated with spirits of ammonia.

Chinese Gold Lacquer.—An imitation of this lacquer

may be prepared by melting two parts of copal and one of shellac until thoroughly mixed, and adding two parts of hot boiled oil. Then remove from the fire and gradually add ten parts of oil of turpentine. To colour, add gum gutta for yellow and dragon's blood for red, dissolved in turpentine.

Preservation of Bronze.—The ugly blackish mass with which most monuments of gun-metal are coated over in such a way as to make them look as if they were made of cast iron, does not consist of sulphuret of copper, as people generally imagine, but of a mixture of coal, sand, etc., with the bronze oxides. The coating referred to cannot be removed either by mechanical means or by washing the surface with reduced sulphuric acid, but an excellent result may be obtained by washing the surface with a concentrated solution of carbonate of ammonia. By this means a coating of platina is formed which prevents the ugly blackish mass heretofore alluded to from making its appearance again on the outside of the monument. It stands to reason that skilled and experienced hands are wanted to perform the operation. If, however, it does not prove successful, Herr Brühl says that Magnus' process might be resorted to, which consists in rubbing the surface of the monument repeatedly, at intervals of several weeks, with a solution of 20 parts of crystallizing acetic acid in 100 parts of neat's-foot oil. By the protoxides of copper and the oleic acid that arise from this, a thin green coating is produced, which prevents dirt and dust from sticking to the surface of the article, and, moreover, involves the formation of a coating of platina, without which bronzes exposed to oxidation in the open air are never likely to remain in a good state of preservation.

Gilding Solution.—A durable and beautiful gilding solution is thus described:—Crystallized phosphate of soda, 60 parts; 10 of bisulphate of soda; 1 of cyanide of potassium; 2½ of chloride of gold; and 1,000 of distilled or rain water—by weight. To prepare this bath properly the water is divided into three portions, viz., one of 700 and two of 150 each. The sodic phosphate is dissolved in the first portion, the chloride of gold in the second, and the bisulphide of soda and cyanide of potassium in the third. The first two portions are gradually mixed together, and the third is afterwards added. With this solution the artisan uses a platinum anode—a wire or strip—adding fresh portions of the gold salt as the solution becomes exhausted.

Substitute for Gold Bronze.—According to experiments, metavanadic acid may be used in the preparation of a substitute for genuine gold bronze. If a solution of sulphate of copper and sal ammoniac is mixed with vanadiate of ammonia and cautiously heated, there is obtained a compound of a splendid gold colour, which is deposited from the liquid in the form of gold-coloured spangles. These readily admit of being ground up with gum and varnishes, cover well, do not change on exposure to the air, and are in every respect equal to gold bronze. Vanadium also yields a series of very fine colours especially adapted for painting on porcelain.

Dead Gilding.—A foreign scientific paper, in an article on colour in electro-gilding, says that a dead gilding will be produced by the addition of a little of the fulminate of gold in solution to the bath immediately before gilding, or by dipping the article (brass or copper) before gilding in a mixture of sulphuric and nitric acids.

To Remove Nickel Plate.—Nickel plating can be removed from brass or iron by placing in a solution of not too dilute hydrochloric acid, more readily in hot than in cold. Hot dilute sulphuric acid dissolves it with some difficulty. Dilute nitric acid operates more quickly.

A Rubber Lubricator for Belts.—Five parts of india-rubber are cut fine and melted together with five parts oil of turpentine, in a well-covered iron vessel, to which are added four parts of resin. This is stirred thoroughly and melted, and four parts of melted wax mixed with the same, the mass being constantly stirred while melting. This composition in its warm state is added, with constant stirring, to a melted mixture of some fifteen parts of fish oil and five parts of tallow, and the whole agitated until it has congealed. The compound is applied to old belts upon both sides in a warm place, and, when the belts are in use, from time to time upon the inner side.

To Soften Rubber Rings.—Rubber rings that have become hard and useless can, it is said, have their elasticity restored by being placed for half an hour in a solution of water and ammonia—about twice as much of the latter as of the former.

Impervious Rubber Packing.—This may be made steam and air-tight by brushing it over with a solution of powdered resin in ten times its weight of strong water of ammonia. At first this solution is a viscid, sticky mass, which, however, after three or four weeks becomes thinner and fit for use. The liquid sticks easily to rubber as well as to wood and metal. It hardens as soon as the ammonia evaporates, and becomes perfectly impervious to liquids.

India-rubber Oil.—One of the recent inventions is an india-rubber oil, designed to protect the surface of iron from rust. To obtain it the rough oils resulting-from the distillation of brown coal, peat, or other bituminous substances are subjected to further distillation; thinly-rolled india-rubber, cut into small strips, is saturated with a four-fold quantity of this oil, and is allowed to stand for eight days ; the mass, thus composed, is subjected to the action of vulcan oil, or a similar liquid, until a homogeneous, clear substance is formed. On the application of this substance—in as thin a form as possible—to a metal surface, it forms, after slow drying, a kind of skin, which, as alleged, insures absolute protection against atmospheric influences, the durability of the covering being also another advantage in its favour.

Using New Pens.—The ink will flow freely from new pens if they are passed before using two or three times through a gas flame to remove the grease with which they are coated.

Indelible Pencils.—Take kaolin, eight parts: finely-powdered manganese dioxide, two parts ; silver nitrate, three parts ; mix and knead intimately with distilled water, five parts. Then dry the mass and enclose it in wood.

Making Rubber Stamps.—Have a vulcanizing apparatus with a thermometer and a lamp under it, such as dentists use ; have an iron printing frame in which you lock up
the type for all the names which you wish to reproduce in
rubber, and of such a size that the plaster mould made from
it can be placed inside the vulcanizer. This mould is made
like an ordinary stereotype mould, by first oiling the type
and then pouring the plaster over it ; when set, take it off
carefully, and do not let it dry, but proceed at once by placing
on top of the mould a piece of sheet-rubber (not pure rubber,
as this is too sticky, but vulcanized, and mingled with sulphur and soap stone). Then have two iron plates, one for
placing on top of the sheet rubber and one below the plaster
mould, and which by proper screws can press together and
squeeze the rubber on the mould. Back up the rubber with a
few sheets of paper, so as to prevent it from sticking at the back
of the iron plate. After screwing down sufficiently, immerse
the mould and rubber in the water in the vulcanizer, screw the
cap on, and heat to 300 deg. Fahr., then let it cool, open the
vulcanizer, take out the mould and rubber, and remove the
rubber carefully. This will be easily done if you have put
the mould, while still wet, in the vulcanizer. Cut up the
rubber so as to separate the various names, glue them to
handles, and your rubber hand-stamps are ready. This is
the regular method, and if you are not satisfied with the impression given it is the fault of the manipulation. Such
stamps give as clear an impression as can possibly be desired
—as clear as metal type. But in printing with them you
must apply a slight pressure only ; the best rubber type can
be made to give bad impressions by defective inking and
rough manipulation in printing.

To Dissolve Rubber.—The best method of dissolving
odds and ends of sheet india-rubber is in a mixture of methylated ether and petroleum spirit or benzoline. The general
method of using up old india-rubber is by heating it with
steam, under which the sulphur discharges, the rubber melts,
runs into hot water, and collects at the bottom of the pit, the
vapour preventing it from burning.

To Vulcanize Rubber.

—The following is a new method of vulcanizing rubber :—For vulcanizing india-rubber there is generally employed a bath of sulphur and steam under pressure. Someone has conceived the idea of substituting a concentrated solution of calcium chloride capable of furnishing a constant temperature of from 150 to 160 degrees. The advantages of this are very important, and may be summed up as follows:—(1) There is no modification required in the apparatus at present existing in factories ; (2) the iron plate vats not being exposed to burning, last much longer ; (3) the capital invested in the bath is insignificant compared with that required by the sulphur bath—sulphur, in fact, costs in Paris about 200 francs per ton, while calcium chloride is but 80 francs ; (4) the daily consumption of the chloride, when the bath has once been prepared, is almost nothing, since this salt is fixed and indecomposable, while with sulphur there is a continual consumption on account of its volatilization and accidental combustion; (5) as a consequence of the suppression of the vapours of sulphur and sulphurous acid in the works the manufacture has no bad influence upon the health of the workmen; (6) there is no danger of fire, and consequently the insurance rates are lower ; (7) the apparatus lasts longer, since the iron employed for locking the moulds no longer has sulphur to combine with and make it brittle ; (8) the consumption of fuel is reduced by about two-thirds, since the bath has a powerful calorific capacity, and the vats may be heated by an open fire ; (9) finally, as the boiling is much gentler and more regular, it gives products of superior quality, and little or no waste, on account of the facility that exists of always keeping the bath at a temperature of between 150 and 155 degrees.

Manufacture of Steel Pens.

—Steel used for making pens reaches the factory in sheets about two feet long by one foot three inches wide, 0.004 inch thick. They are cut into bands of different widths, according to the dimensions of the pen required, the most usual widths being two, two and one-half, and three inches. The bands are then heated in an iron box and annealed, when they are passed on to the rolls and reduced to the desired thickness of the finished pen, thus being transformed into ribbons of great delicacy, about four feet long. The blanks are then stamped out from the ribbons by a punching machine, the tool of which has the form of the pen required. The blanks leave the die at the lower part of the machine, and fall into a drawer with the points already formed. They are then punched with the small hole which terminates the slit, and prevents it from extending, and afterward raised to a cherry-red heat in sheet-iron boxes. The blanks are then curved between two dies, the concave one fixed and the convex brought down upon it by mechanism. The pens, now finished as regards their form, are hardened by being plunged, hot, into oil, when they are as brittle as glass. After cleansing, by being placed in a revolving barrel with sawdust, they are tempered in a hollow cylinder of sheet iron, which revolves over a coke fire after the manner of a coffee roaster. The cylinder is open at one end, and while it is being turned, a workman throws in twenty-five gross of pens at a time, and watches carefully the effect of the heat on the colour of the pens. When they assume a fine blue tint, he pours the pens into a large metal basin, separating them from one another, to facilitate the cooling. After this process, which requires great skill and experience, comes the polishing, which is effected in receptacles containing a mixture of soft sand and hydrochloric acid, and made to revolve. This operation lasts twenty-four hours, and gives the pens a steel-grey tint. The end of the pen, between the hole and the point, is then ground with an emery wheel, revolving very rapidly. There only now remains to split the pens, which is the most important operation, being performed by a kind of shears. The lower blade is fixed, and the upper one comes down, with a rapid motion, slightly below the edge of the fixed blade. To give perfect smoothness to the slit, and at the same time make the pens bright, they are subjected to the operation of burnishing by being placed in a revolving barrel almost entirely filled with boxwood sawdust.

To Render India-rubber Gas Tubes Odourless.—

It is well known that all rubber tubing soon becomes disagreeable from the gas that slowly permeates it and escapes from its surface. The following is given by a German scientific journal as a remedy for the disagreeable property of this very useful adjunct to portable gas-lamps. The method consists in mixing ordinary alcohol of about 36 per cent. with an equal volume of genuine linseed oil, shaking well until they are intimately mixed, and then applying the mixture to the tube. This is done by pouring a few drops upon a small rag and rubbing it upon the tube, which should be moderately stretched. and rubbed until it is quite dry, which takes but a few minutes. The oily mixture is to be applied three or four times at intervals of a few days. This treatment is said to render the tube perfectly gas-tight and odourless without destroying its elasticity or colour.

Plumbago in Lead Pencils.—The plumbago used in the manufacture of lead pencils is made into a hard form by mixing with clay in various proportions according to the hardness required, and baking in an oven.

How to Make Metallic Writing Pencils.—These are made of an alloy of lead, bismuth, and quicksilver. The ingredients vary according to the desired degree of hardness. The ordinary proportions are : lead 70, bismuth 90, and quicksilver 8 parts by weight. A larger proportion of lead and quicksilver makes the pencil softer, and produces darker marks in writing. The lead and bismuth are melted together and allowed to cool somewhat, when the quicksilver is added, and the composition cast in proper moulds.

A Home-made Fountain Pen.—Take two ordinary steel pens of the same pattern and insert them in the common holder. The inner pen will be the writing pen. Between this and the outer pen will be held a supply of ink, when they are once dipped into the inkstand, that will last to write several pages of manuscript. It is not necessary that the points of the two pens should be very near together, but if the flow of ink is not rapid enough the points may be brought nearer by a small piece of thread or a minute rubber band.

Rubber-Stamp Moulds.—Moulds for rubber stamps are made of plaster of Paris. The rubber is pressed in with a small press or clamp, then placed in a small vulcanizing oven heated by steam or a furnace to a temperature from about 250 to 275 degrees.

Colouring of Pencils.—The colouring matter of crayon pencils is derived from various sources. Red is mineral red chalk, and blue is soluble Prussian blue with whiting, and so on according to the tint or colour required. You can buy them much cheaper than you can make for yourself.

To Render Pencil Notes Indelible.—Pencil notes found in a book, or placed there as annotations, may be rendered indelible by washing them with a soft sponge dipped in warm vellum size or milk.

Counting Lead Pencils.—In factories where pencils are made in numbers, a simple method of counting has been devised, with a view to saving time and trouble. Strips of wood are employed, having in each 144 grooves, and the workmen, taking up a handful of pencils, rapidly rub them along the board once and back, thus filling all the grooves in which the pencils lie, similarly to pens on a rack. In five seconds a gross may thus be counted, without the least likelihood of making a mistake, and much time and labour are saved.

Composition of Coloured Pencils.—The first four grades are as follows :—

	No. 1. Very soft.	No. 2. Soft.	No. 3. Hard.	No. 4. Very hard.
Aniline	50 parts.	46 parts.	30 parts.	25 parts.
Graphite	37.5 „	34 „	30 „	25 „
Kaolin	12.5 „	24 „	50 „	50 „

For purple an aniline violet is used ; for other colours various shades of aniline. The cheaper qualities of coloured pencils consist simply of the colouring material mixed with kaolin or clay.

How Common Crayons are Made.

—The process of manufacture of the common chalk or school crayon consists of mixing equal parts of washed pipeclay and washed chalk into a paste with sweet ale made hot, into which a chip or two of isinglass has been dissolved. This paste is rolled out with a rolling pin, cut into slips, then rolled into cylinders by means of a small piece of flat wood, cut into lengths, and finally placed in a slow oven or drying stove until hard.

Pencils for Glass and Porcelain.

—Pencils for these purposes are produced in the following manner :—
1. Black pencils—10 parts of the finest lampblack, 40 parts of white wax, and 10 parts of tallow. 2. White pencils—40 parts of Kremser white, 20 parts of white wax, and 10 parts of tallow. 3. Light blue pencils—10 parts of Prussian blue, 20 parts of white wax, and 10 parts of tallow. 4. Dark blue pencils—15 parts of Prussian blue, 5 parts of white wax, and 10 parts of tallow. 5. Yellow pencils—10 parts of chrome yellow, 20 parts of white wax, and 10 parts of tallow. The colour is mixed with the body of the wax and tallow warm, triturated, exposed to the air for drying, so that the mass can be pressed by means of a hydraulic press into round pencils in the same way as lead pencils are formed. The pencils are dried after pressing by exposing them to the air, until they have the proper consistency, and are then glued into the wood.

The Lead Pencil.

—There is no pencil lead, and there has been none for fifty years. There was a time when a spiracle of lead, cut from the bar or sheet, sufficed to make marks on white paper, or some rougher abrading material. The name of lead pencil came from the old notion that the products of the Cumberland mines were lead, instead of being plumbago, or graphite, a carbonate of iron capable of leaving a lead-coloured mark. With the original lead pencil or slip, and with the earlier styles of the " lead " pencil made direct from the Cumberland mine, the wetting of the pencil was a preliminary to writing. But since it has become a manufacture the lead pencil is adapted, by numbers or letters, to each particular design. There are grades of hardness, from the pencil that may be sharpened to a needle point, to one that makes a broad mark. Between the two extremes there are a number of graduations that cover all the conveniences of the lead pencil. These graduations are made by taking the original carbonate and grinding it, and mixing it with a fine quality of clay in differing proportions, regard being had to the use of the pencil. The mixture is thorough, the mass is squeezed through dies to form and size it, is dried, and incased in its wood envelope.

Composition for Crayons.—A newly patented composition consists of:—Water, 8 lbs.; kaolin, 15 lbs.; wheat-flour, 1 lb.; soapstone, 1 lb.; Paris white, 45 lbs. A thick paste is made of flour and water, which is dissolved in 8 lbs. of warm water, the other ingredients being thoroughly mixed therein by agitation. The water is pressed out of this composition, which is squeezed through dies of suitable shape. The crayons harden by exposure to the air. This composition makes a crayon that does not require heat to harden, is free from dust, white in colour, and, by varying the proportions, can be made in colours by colouring the water.

Transferring Design for Embroidery.—A method has been devised for transferring a design on a tissue to be embroidered. A mixture is made of printers' ink, glycerine, and wax, and the design printed in it on a sheet of paper which has been folded, a solution of stearine and wax having been applied between the folds. The paper is then placed over the material upon which the design is to be transferred, and is pressed by means of a smoothing bone.

Oil Painting on Terra Cotta.—Now that decorative art is much patronized, the following may prove interesting to amateurs:—Upon terra cotta of a light tint the design is drawn with a lead pencil; upon that of a dark tint use the coloured impression paper. Place the article between piles of books, or fill a box with sand, and lay or stand it into this in the position required; see that the right arm rests upon an even plane with the article to be decorated. A

terra-cotta medium is made from a small quantity of gum
arabic dissolved in water, to which is added a little syrup ;
go over the entire article with a flat brush dipped into the
medium ; when dry, repeat the wash. When dry, the article
is ready for the oil colours. Mix these with flake white. Lay
the colours on fairly thick and let them dry for some hours,
then tint and finish with the colours necessary without the
flake white. When finished and quite dry, varnish with copal
or mastic. The artist should have at hand two or three fine oil
brushes and a flat brush, and the necessary colours. Those
being indispensable are the following : black, burnt light ochre,
terra di sienna, Indian red, and flake white. The artist is re-
minded that vases of antique shape look best when decorated in
antique designs.

Enamel for Iron.—A good enamel coating for cast
iron, wrought iron, or steel, one that will not crack on being
subjected to moderate changes of temperature, has long been
a desideratum, and this is now claimed to have been discovered. In the case of an opaque enamel being required,
as, for instance, a basis for vitrified photographs, about 8 parts
of oxide of tin are to be added. About 125 parts by weight
of ordinary flint-glass fragments, 20 parts of carbonate of
soda, and 12 parts of boracic acid are melted together, and
the fused mass poured out on some cold surface, as of stone
or metal. When this has cooled sufficiently it is pulverized,
and a mixture made of the powder, together with a silicate of
soda of 50° R. With this substance the metal is glazed and
heated in a muffle or other furnace until it is fused. This is
said to prove an effective application for the purpose as compared with other methods, and at least possesses the great
advantage of simplicity.

To Protect Brass against Acids.—Melted paraffin
wax will effectually protect brass from the action of nitric acid.

Taking Out Writing Ink.—Common writing ink may
be removed from paper without injury to the print by oxalic
acid and lime, carefully washing it in water before restoring
it to the volume.

Modelling Wax.—This wax, such as sculptors sometimes use for modelling small figures, etc., is made of white wax melted and mixed with lard to make it workable. In working it the tools used, the board or stone, are moistened with water to prevent its adhering: it may be coloured to any desired tint with a dry colour.

Crystallotypy.—A novel process has been published of producing artificial tint plates. The name "Crystallotypy" is given because the production of the artificial plates depends upon a crystallizing of certain salts, poured in a dissolved condition upon a level glass plate. By it are secured designs like frost work seen in winter upon the window panes, and if impressions are made to transfer paper with a type-printer's roller such designs may be transferred to stone. The following recipe is given:—Take either 200 parts of water and 20 parts of sulphate of copper, or same proportions of either water and red prussiate of potassa, or of water and alum, or, finally, of water and bichromate of potassa, cover with the solution a level and perfectly well-cleaned glass plate, and a few hours are sufficient to cause the water to evaporate and the design to appear. Instructions are also given how electrotypes may be made from such plates. The idea is certainly worthy of notice.

Artificial Dextrine.—Two parts nitric acid to 300 parts of water, mix with 1,000 parts of dry starch. This mixture is then subjected to heat. It may also be produced by heating starch with diastase, a peculiar azotized substance contained in malt, which effects the conversion of starch, first into dextrine, then into grape sugar.

Colouring Photographs.—A new process has been patented. This consists of immersing the photographs in a solution of naphtha, paraffin, mastic drops, ether, and vinegar, and applying to the back in oil paint the desired shade and tone, and also applying a mixture of glue and glycerine to the back, and pressing the back to canvas until cohesion takes place, whereby the whole picture will be flexible, and have the appearance of having been painted on canvas.

Prevention of Frost on Glass.

Prevention of Frost on Glass.—A good many suggestions, all more or less practicable, have been made how to prevent frost on windows. A very thin coat of glycerine applied on both sides of window glass will prevent any moisture forming thereon, and will stay until it collects so much dust that you cannot see through it ; for this reason it should be put on very thin. If used on a looking-glass you can shave yourself in an ice-house and the glass will not show your breath. Doctors and dentists use it on small glasses with which they examine the teeth and throat. Surveyors use it on their instruments in foggy weather, and there is no film to obstruct the sight. Locomotive engineers have used it as a preventive of the formation of frost. In fact, it can be used anywhere to prevent moisture from forming on anything. It does not injure the usefulness of field glasses, etc.

Waterproofing Material.

Waterproofing Material.—Basic silicate of alumina is said to be far better than tungstate of soda, because it does not require, like the latter, frequent renewals.

Imitation Ivory Casts.

Imitation Ivory Casts.—A foreign scientific paper says that by painting a plaster cast with yellow beeswax dissolved in turpentine it will in a short time be hardly distinguishable from ivory.

Imitation of Ground Glass.

Imitation of Ground Glass.—A good result may be obtained by brushing (dabbing) the glass lightly with a solution obtained by boiling a teaspoonful of rice in a pint of water for half an hour.

Another Method of Imitating Ground Glass.

Another Method of Imitating Ground Glass.—This can be produced by dissolving about two tablespoonfuls of Epsom salts in a pint of ale and painting the glass with the mixture. When it is dry the glass will appear as if frosted.

Cloth Tracings.—A correspondent of an industrial journal refers to the difficulties encountered in tracing upon cloth or calico, especially the difficulty of making it take the ink. In the first place the tracing should be made in a warm room or the cloth will expand and become flabby. The excess of glaze may be removed by rubbing the surface with chamois leather on which a little powdered chalk has been strewn, but this practice possesses the disadvantage of thickening the ink, besides making scratches which detract from the effect of the tracing. The use of ox-gall, which makes the ink " take," has also the disadvantage of frequently making it " run," while it also changes the tint of the colours. The following is the process recommended :—Ox-gall is filtered through a filter-paper arranged over a funnel, boiled, and strained through fine linen, which arrests the scum and other impurities. It is then placed again on the fire, and powdered chalk is added. When the effervescence ceases the mixture is again filtered, affording a bright colourless liquid if the operation has been carefully performed. A drop or two must be mixed with the Indian ink. It also has the property of effacing lead-pencil marks. When the cloth tracings have to be heliographed raw sienna is also added to the ink, as this colour unites with it the most intimately of any, besides intercepting the greatest amount of light.

Leaf Photographs.—A very pretty amusement is the taking of leaf photographs. One process is this :—Put an ounce of bichromate of potassium into a pint bottle of water. When the solution has become dissolved, pour off some of the clear liquid into a shallow dish; on this float a piece of ordinary writing paper till it is thoroughly moistened. Let it become dry in the dark ; it should be of a bright yellow. On this put the leaf—under it a piece of soft black cloth and several sheets of newspaper. Put these between two pieces of glass (all the pieces should be of the same size), and with spring clothes-pins fasten them together. Expose to a bright sun, placing the leaf so that the rays will fall as nearly perpendicular as possible. In a few moments it will begin to turn brown, but it requires from half an hour to several hours to produce a perfect print. When it has become dark enough, take it from the frame and put it in clear water, which must be changed every few minutes until the yellow part becomes white. Sometimes the leaf veinings will be quite distinct. By following these directions it is scarcely possible to fail, and a little practice will make perfect.

Translucent Pictures.—A new method of making "linophanies," or translucent pictures, comprises the following processes :—Paper pulp, white or coloured, in a liquid or semi-liquid state, is poured into a mould, usually of metal, the bottom of which is engraved to form or produce the desired design or picture. Enough pulp must be poured in to leave a thin film of paper over the highest lines of the engraved surface, when the pulp is dried, as will be hereinafter described. Those portions of the engraving which correspond to the dark shades of the picture are cut deeper than those which correspond to the lights, the depth varying with the depth of the shade. Upon the paper pulp in the mould is spread a piece of gauze, fine linen, or other similar material that will not adhere to the pulp, but will permit the passage of water, and on this is placed blotting paper in one or more layers, the whole being subjected to pressure in an ordinary press. The blotting paper thus absorbs the greater portion of the water from the pulp, and the latter is pressed into all the finer lines of the engraving. For the blotting paper any other absorbent material—as some kinds of felt—may be substituted. After removing the mould from the press the blotting paper and the linen or gauze separating material are removed and the mould containing the partially dried pulp is subjected to artificial heat, as in a stove or kiln, to dry out the remaining moisture. The dried paper pulp is now removed from the mould, and will be found to consist of a continuous imperforated paper leaf bearing the design or picture, which will be fully brought out when the sheet is held up between the eye and a strong light. The thinner portions of the sheet will represent the lights of the picture, and the thicker (and less translucent) portions will represent the shades. The picture or ornament thus produced is called a "linophany,"

U

and may be employed for lamp or gas shades, for transparent pictures for windows, etc., or, indeed, for any ornamental purposes to which such a picture or design is adapted. In lieu of using artificial heat to dry the pulp in the mould it may be dried by the natural evaporation of its moisture. The mould is usually engraved, but the design may be formed in any way—as by pressure from a hardened relief-plate.

How to Preserve Pencil Sketches.—The pencil drawings of mechanical draughtsmen and engineers may be rendered ineffaceable by a very simple process. Slightly warm a sheet of ordinary drawing paper, then place it carefully on the surface of a solution of white resin in alcohol, leaving it there long enough to be thoroughly moistened ; afterwards dry it in a current of warm air. Paper prepared in this way has a very smooth surface, and in order to fix the drawing the paper is to be simply warmed for a few minutes. This process may prove useful for the preservation of plants or designs when the want of time or any other cause will not allow of the draughtsman reproducing them in ink. A simpler method than the above, however, is to brush over the back of the paper containing the charcoal or pencil sketch with a weak solution of white shellac in alcohol.

Photographic Printing in Colours.—In this process it is necessary to use *coloured negatives*—that is, ordinary negatives that have been hand-painted in the required tints with transparent colours. (1) Take a piece of ordinary sensitized paper, and wash it to remove any free silver nitrate. (2) Place the washed paper in a solution of protochloride of tin, and expose to weak light until the silver chloride is reduced to subchloride, and the paper assumes a uniform grey colour. (3) Float the paper in a mixed solution of chromate of potash and sulphate of copper, and dry in the dark. The paper is now sensitive to all the colours of the spectrum, and by printing on it with a coloured negative the colours of the negative will be reproduced. After printing, wash with cold water, and dry.

Spatterwork.—A new and somewhat attractive sort of ornamental work is to be seen in some stationers' shops. The materials are delicately formed leaves, as vines, ferns, rose, geranium, or maple, and ornamental letters cut from printed matter. Press the leaves carefully till quite flat. For implements all that are needed are an old tooth-brush, a flat board, and white or tinted paper. Arrange your pressed leaves or letters on your paper, fastening down firmly with fine needles, which wipe after using, dip the brush in the dye, and, holding a sieve over your pattern, rub carefully at first over the wires, making it heavy or light as desired. Let the leaves and letters dry, then remove, and they can be used again. That is all. Anyone who has not seen this work will be surprised at the beautiful effects from so simple a process. For working with paper, use any colours that will not dry and rub off. Beautiful crosses, shaded and twined with pressed vine, with ferns at the base, can be easily made after a little practice ; also wall-pockets, letter-holders, and comb-cases. Old cigar-boxes, covered neatly and carefully spattered, are very ornamental.

A New Method of Mounting Prints on Cloth.

—Prepare several yards of cloth at a time by sizing with starch, and always keep a roll of it on hand ready for use. While damp the cloth is stretched (not too tightly) on a frame, and sized plentifully with warm starch paste, made rather thin and spread on evenly. When dry cloth is cut to the size required before mounting (allowance being made for the expansion of the prints), if the starch for mounting be used while warm (which is preferable), it should be as stiff as can be conveniently spread on the print, for the reason that it will expand the cloth less and dry quicker. From the moment the first print touches the cloth despatch is important; therefore both prints are first pasted, one being laid aside ready to be picked up quickly. The first print is rubbed down more expeditiously with a hand-roller than with the hands. When the second print is properly laid on there is less occasion for haste, and rubbing down by hand is preferable, because, although the roller does the work perfectly on the first print mounted, it is liable to leave air-bubbles in rolling down the second one. To avoid bubbles the hand-rubbing should be towards the middle of the print, and not in every direction from the centre. When the mounting is completed, the prints are placed between papers and covered immediately with several folds of cloth of sufficient weight to keep them in place. To facilitate drying they may be aired after an hour or two, and placed between dry papers, and again covered with the cloth.

Ormolu.—The ormolu of the brass founder, popularly known as an imitation of red gold, which is extensively used by French workers in metals for artistic and decorative work, is composed of copper and zinc. A greater proportion of copper and less zinc is used in its preparation than in making ordinary brass. It can be readily cleansed with acid, and can be burnished with facility. To give this material a rich appearance it is not unfrequently brightened up after " dipping "—that is, cleansing in acid—by means of a scratch brush (a brush made of fine brass wire), the action of which helps to produce a brilliant gold-like surface. It is protected from tarnish by the application of lacquer.

Cleaning Powder for Show-Windows.—A good one, which leaves no dirt in the joints, is prepared by moistening calcined magnesia with pure benzine, so that a mass is formed sufficiently moist to let a drop form when pressed. The mixture has to be pressed in glass bottles with ground stoppers, in order to retain the easily volatile benzine. A little of the mixture is placed on a wad of cotton and applied to the glass plate. It may also be used for cleaning mirrors.

How to Bronze a Plaster Cast.—A scientific contemporary says the best way is to go over the figure with isinglass size until it holds wet, or without any part of its surface becoming dry; then with a brush go over the whole, taking care to remove, while it is yet soft, any of the size that may lodge on the delicate parts of the figure. When it is dry take a little thin oil gold size, and with as much as just damps the brush go over the figure with it, allowing no more to remain than causes it to shine. Set it aside in a dry place free from smoke, and in forty-eight hours the figure is prepared to receive the bronze. After having touched over the whole figure with the bronze powder, let it stand another day, and then with a soft dry brush rub off all the loose powder, particularly from the points or from the more prominent parts of the figure.

How to Clean Chamois Leather.—When a chamois skin gets into a dirty condition, rub plenty of soft soap into it, and allow it to soak for a couple of hours in a weak solution of soda and water, rubbing it till it appears quite clean. Now take a weak solution of warm water, soda, and yellow soap, and rinse the leather in this liquor, afterwards wringing it in a rough towel, and drying it as quickly as possible. Do not use water alone, as that would harden the leather and make it useless. When dry, brush it well and pull it about; the result will be that the leather will become almost as soft as fine silk, and will be to all intents and purposes far superior to most new leathers.

Lustra Painting.—Lustra painting is an invention for household decoration, lately introduced by an artist of repute, and belongs, by reason of its facility of manipulation and its demands upon refined and delicate taste, to feminine hands. It may be used for almost everything susceptible of ornamentation—altar cloths, ladies' dresses—and takes the place of expensive embroidery, at only a fraction of its time and cost. It can be applied to every fabric from velvet to linen—for curtains, screens, portières, dados, friezes, wall panels, and ladies' flounces, and also to wood and the various articles made of terra cotta. The colours are prepared from metallic bases, but the fabrication of them is a secret resting with the inventor. The lustra colours are all dry and are mixed on a peculiar palette with little saucer-like hollows to hold them in fluid form. They are used with a colourless medium, a bottle of which accompanies each box.

The material requires no preparation.   The colours once laid on are firm.   Used on linen it may be washed, always, however, on the reverse side, and with brushing instead of rubbing.   This art has the advantage over needlework of being infinitely quicker of execution and far less expensive in material.

"Graph" Process with Printers' Ink.—Make the ordinary pad with glue and glycerine, only using a larger proportion of gelatine.   For ink take a concentrated solution of alum coloured with any aniline dye to render the marks visible, and execute the drawing or writing on good stiff paper.   Then wipe the pad with a damp sponge, and allow the moisture to remain a few minutes to make the surface sufficiently absorbent.   Now place the paper, face down, on the pad.   After a few minutes remove the paper, and the writing or drawing is transferred to the pad ready for printing.   This is done by applying printing ink to the surface by means of an india-rubber roller, the ink adhering only to the lines made by the alum solution.   To take the impression, place upon the pad a sheet of paper, slightly damped, and gently rub it with the fingers.   Remove the paper, which is now printed, ink the pad again, and keep on in the same way as long as the impression is sharp enough.   A large number of copies may thus be taken, of a good black, and as permanent as ordinary printing.

Cheap and Effective Window Signs.—A glass window or door may be very effectively lettered at a merely nominal cost in the following manner :—The wording is to be set up in type in the style and of the size required, pulled, and a rough impression affixed to the outside of the glass.   From a second impression the letters are separately cut out and carefully pasted on the inside glass, using the outside and loosely-affixed impression as a guide for position.   If air-bubbles form they must be carefully rubbed out.   If difficult to get rid of, a pin-prick will let out the air, and a little pressure while the paste is still wet will cause them to entirely disappear.   When perfectly dry the pasted-on letters are to receive a coat of ordinary size, and to make quite certain of good work a second coat should be applied after the first has thoroughly dried. The paste and size that have splurged over and dried on the glass are to be carefully removed with a damp cloth. Now get a chemist or oilman to rub down some oxide of zinc in some linseed oil in such proportions that the mixture will assume the consistency of cream. Paint the mixture on the inside of the glass over the entire surface, including the backs of the letters. If it dries streaky a second coat may be applied after the work is thoroughly set, when the streaks will completely disappear. As the impressions were originally printed, so the sign will show in black or red, or both, and the groundwork will appear of a French grey. Lettering done in this manner will last a century.

Tinting Maps.—Maps, as well as architects' and engineers' designs, plans, sections, drawings, etc., may be tinted with any of the simple liquid colours, preference being given to the most transparent ones, which will not obscure the lines beneath them. To prevent the colours from sinking and spreading, which they usually do on common paper, the paper should be wetted two or three times with a sponge dipped in alum water (three or four ounces to the pint), or with a solution of white size, observing to dry it carefully after each coat. This tends to give lustre and beauty to the colours ; the colours for this purpose should also be thickened with a little gum water. Before varnishing, two or three coats of clean size should be applied with a soft brush, the first one to the back.

Photo-Zincotypes in Colours.—A new process for coloured prints, called "photo-chromotypes," has been perfected by Angerer & Göschl, of Vienna. The principle is similar to the coloured "lichtdruck." First, photo-lithographs are made from the picture to be multiplied, which serve to some extent as copies for the draughtsman. The latter works up first only such parts as are to be yellow; upon a second sheet those only intended to be blue, and so on. Negatives are produced which show only a picture of the blue parts, others for yellow, red, etc. From these negatives zinc printing-plates are etched in half-tone, and the rest of the manipulation is the same as the fitting of the several colour stones in chromo-lithography.

To Remove Ink-Spots from Cloth, etc.—The chemical agents generally used are chloride of lime and oxalic acid, but these are not without danger to delicate fabrics. A better detergent is a lye of pyro-phosphate of sodium. There is a proceeding still more simple, which only demands a little care, but which can be employed successfully in materials of fine texture and faint colours. Drop a little melted tallow on the spot of ink, before putting the article into the wash-tub, and then add a good lye of soda. In attacking the grease the lye will also remove the ink.

Removal of Grease Spots from Engravings and Books.—Grease spots, if old, may be removed by applying a solution of varying strength of caustic potash upon the back of the leaf. The printing, which looks somewhat faded after the removal of the spot, may be freshened up by the application of a mixture of 1 part of muriatic acid and 25 parts of water. In the case of fresh grease spots, carbonate of potassa (1 part to 30 parts of water), chloroform, ether, or benzine, renders good service. Wax disappears if, after saturating with benzine or turpentine, it is covered with folded blotting paper, and a hot flat-iron put upon it. Paraffin is removed by boiling water or hot spirits. Ink spots or rust yield to oxalic acid in combination with hot water ; chloride of gold or silver spots, to a weak solution of corrosive sublimate or cyanide of potassium. Sealing wax is dissolved by hot spirits, and then rubbed off with ossa sepia. Indian ink is slightly brushed over with oil, and after twelve hours saponified with salmiac ; any particles of colour still remaining must be removed with rubber. Blood stains disappear after the application for twenty minutes of chloride of lime ; the yellowish stain still remaining yields to a weak acid. Fresh spots of paste are removed with a moist sponge, older ones with hot water. Fusty stains of yellowish colour surrounded with a darker line disappear if the paper is bathed in clean water to which some chloride of lime has been added. If they are found in bound books, linen damped in the same liquid is placed on both sides of the discoloured leaves, whilst the latter are separated from the other leaves by tinfoil. As soon as the spots have disappeared, the linen and tinfoil are removed, the leaves placed between blotting paper, and the book is closed. If there are many fusty spots in the book, the binding is taken off, and the whole volume placed for a night in chloric water. The separated parts are then hung up to dry, and the book freshly bound. If the spots are large, and dotted with black points, tartaric acid is applied.

Flower Barometers.—To prepare paper flowers so that the atmosphere will change them to different colours, in such a manner that they can be used as a barometer, the paper is saturated with a moderately concentrated aqueous solution of cobalt-chloride ; press and dry. When properly prepared, dry air develops a blue colour, and moist air a pink tint. The flower barometer does not foretell the weather, but simply indicates the hygroscopic condition, or the amount of moisture contained in the surrounding air.

Hints to Draughtsmen.—In mixing up inks the process is very much expedited by heating the dish and water in which it is mixed before commencing. It often happens in the summer that the flies walk over a tracing and eat off the ink in a very provoking manner. The use of vinegar instead of water will prevent this. In making a tracing the cloth will take the ink much better if it is rubbed over with chalk. Tracing cloth that has been rolled up may be straightened out effectually and expeditiously by drawing it over the edge of a table or drawing-board, holding it down meantime with an ordinary three-cornered scale. When there are a large number of drawings made and kept, a great deal of trouble and confusion can be avoided by making all the drawings on extra standard sizes. If a size of 16 × 24 in. be adopted, then the next larger size would be equal to two of these, or

24 × 32 in. This enlarging or reducing process may be carried as far as the circumstances require, but it is always best to do it by the doubling or halving process if possible. One of the advantages of standard sizes of drawings is that they may be kept in a case of drawers, the size of which is made to accommodate the standard sizes determined upon.

To Clean Gilt Frames.—Use a soft sponge moderately moistened with spirits of wine ; allow to dry by evaporation. Do not use a cloth, and avoid friction. Another way is to use a very soft shaving brush, and to gently rub backward and forward a lather of curd soap. Rinse with water at about blood heat. This applied morning after morning to old and dirt-covered oil paintings will greatly restore them. In adopting this plan with regard to gilt frames around water-colours or prints, be sure that not enough moisture is used to run off the frame, or the paper will be stained. The cleaning applies to gold frames only. Dutch metal will bear no cleaning, but a new material, not absolutely gold, but very like it, will stand any amount of soap and water.

Wood Moulding for Picture Frames.—These are cut in a machine, brushed over with plaster of Paris, and smoothed down with a steel tool of the same form as the moulding. The plaster has a little glue mixed with it to make it adhere.

A New Copying Process.—A Hungarian chemist has patented a process by which a paper impervious to water is painted with the following solution :—Gelatine, 1 part ; glycerine, 5 parts; Chinese gelatine, 0.2 parts; water, 1 part. The manuscript is written with the following solution:—Water, 100 parts ; chrome alum, 10 parts ; sulphuric acid, 5 parts; gum arabic, 10 parts. The manuscript is laid on the first paper, and the latter is thereby rendered incapable of taking up an aniline colour solution with which the surface is then flowed. Excess of colour is absorbed with silk paper, and negative impressions taken on clean paper.

Removal of Gum Water Stains.—The greasy-looking stain or blot left by gum water when used on paper from which the size has been removed, or which otherwise has become more than usually permeable, may be removed by the addition of sulphate of alumina to the mucilage. One formula runs—2 grammes of crystallized sulphate of alumina dissolved in 20 of water is added to 250 grammes of gum arabic (2 to 5 strength). Another formula gives exactly twice as much sulphate as above; in another the gum mucilage is directed to be made of greater strength, and the alumina salt to be added in the proportion of one thirty-third part of the weight of the gum employed.

Rusting of Screws.—Printers and their allies will be glad to know how to prevent screws from becoming fixed from rust. It is well known that iron screws are very liable to rust, more especially when they are placed in damp situations. When employed to join parts of machinery they often become so tightly fixed that they can only be withdrawn with considerable trouble—a fracture sometimes resulting. In order to avoid this inconvenience screws are generally oiled before being put in their places, but this is found to be insufficient. A mixture of oil and graphite will effectually prevent screws becoming fixed, and, moreover, protect them for years against rust. The mixture facilitates tightening up, is an excellent lubricant, and reduces the friction of the screw in its socket.

To Cleanse Old Engravings.—If brown spots and rings of mildew have not made their appearance, float the engraving, face downward, for twenty-four hours on a large quantity of water in a vessel free from grease. Lift it from the water on a clean sheet of glass, drain, transfer to blotting paper, dry, rub with bread, and iron. If the stains are not removed by this plan, place the engravings in a shallow dish and pour water over them until they are completely soaked. Carefully pour off the water, and replace it by a solution of chloride of lime (one part of liquor calcis chloratæ to thirty-nine parts of water). As a rule, the stains disappear. If not, pour on the spot pure liquor calcis chloratæ. As soon

as the stain disappears, carefully wash the engraving with clean water till all the chlorine is removed. Then steep it in a weak solution of glue or gelatine. Dry between blotting paper under a weight, and iron with a sheet of clean paper between the iron and the print. Small grease spots may be removed by putting powdered French chalk over them, a piece of clean blotting paper over the chalk, and a hot iron over that. If the stains are larger, benzine may be used, applying it round the stain, before touching the stain itself.

Albertype.—In the Lichtdruck process (named after its inventor) glass is substituted as the basis for the gelatine, because it gives the opportunity of hardening the film from the back by exposure to light. By a preliminary coating of bichromatized albumen, also hardened by exposure from the back, this is effected more completely. Lichtdruck prints are of very great beauty, and hardly distinguishable at a little distance from silver prints from the same negative.

Faded Goods in Shop Windows.—It is stated that plush goods and all articles dyed with aniline colours, faded from exposure to the light, will look as bright as ever after being sponged with chloroform. The commercial chloroform will answer the purpose very well and is less expensive than the purified.

Phytochromotypy.—This is a process of producing impressions of leaves and plants, and is effected as follows : The plant is first dried and flattened by pressure between un-sized paper, or it may be done rapidly with a hot iron. The surface to be copied is then brushed with a solution of aniline colour in alcohol and allowed to dry, which will take place very rapidly. If the impression is to be taken on paper, im-merse the latter in water for a few seconds, and remove the excess by pressing between blotting paper. Place it then on some non-absorbent surface, and apply the plant coloured side down; place over it a sheet of strong paper, and while it is held securely in position stamp the whole surface with a wad of cotton. A cold iron may be lightly passed over the paper instead of using the cotton, and if a few sheets of tissue paper are interposed between the paper and the plant, its outlines and veins principally will be copied, while without it the whole surface may be impressed on the paper. If the paper which is to receive the impression is moistened with alcohol instead of water, the impression will be brighter, and the paper will retain its lustre or glaze better. If a very light coating of glycerine be spread upon the coloured plant when perfectly dry, and the excess removed by unglazed paper, one or more prints may be immediately taken upon dry paper or other dry surface. If the print shows blots when a strong colour is used, pass over the surface with a pencil wet with a solution of saltpetre, which will moderate the impression. Different parts of the plant may be coloured differently to conform to nature or individual taste. Defects may be touched up with a pen dipped in the colour.

Indigo Printing on Cloth.—Having soaked the indigo in caustic soda lye, it is ground, then mixed with water and dextrine to a paste, and treated with dry caustic soda, taking care that the temperature does not rise above 25° R. Cloth mordanted with grape sugar is printed with this mixture, dried and steamed for about five minutes in a continuous steaming apparatus, whereby the indigo is reduced. In the subsequent passage through water the indigo is oxidized for 30 minutes, and the cloth dried. For 2,000 g. indigo, 400 g. caustic soda, 500 g. dextrine, 6,800 g. water, and 26 g. mordant for cloth are required.

Ruling Tissue Paper.—This is a process for ruling light papers (such as tissue paper) which cannot be properly carried through and ruled upon the ordinary ruling machines on account of its liability to crimp or gather under the pens, thus causing irregular or imperfect lines. By the use of this process it is intended also to so rule the paper that the lines as they are drawn may strike through the sub-stance of the paper so as to present both surfaces of the finished sheet alike, and further, to rule the paper in a con-tinuous sheet as it is delivered from the paper-making machine hereinafter specified. The paper is drawn directly, under tension, over a roller or mandril, and from thence over a felt-covered roller, which revolves in a trough containing a solution of saccharine matter and ox-gall, by which it is moistened, so as to put it in proper condition to receive the ink. From the roller it passes under a brush-roller, having its periphery armed with bristles and travelling in a direction opposite to the line of travel of the paper. The office of this roller is to smooth out the wrinkles from the paper caused by the moistening operation. The paper then passes to a drying-roller, and, on its way there, to below the ruling-pens, which are supported and supplied with ink in the same manner as in an ordinary paper-ruling machine. The paper thus moistened and prepared receives the ink-lines while in such condition that the same may strike through its body. The drying-roller is suitably heated, preferably by means of steam, so as to dry the ink as soon as deposited from the pens, and thus prevent it from spreading and blurring the lines. From the drying-roller the paper passes between calendering rolls, one of which is suitably heated, so as to finally dry and surface the paper. The paper is finally wound upon a drum, which serves not only to roll and store it, but to draw it and keep it at the proper tension under the pens, so as to receive the lines properly, the travel of the web or blanket being so regulated with respect to the rate of rotation of the drum that the sheet will always be taut or " stretched " when passing under the pens.

Printing Stained Glass.—Something very like stained glass to the eye, and as pretty to look upon as the genuine vitreous article of many colours, is now produced abundantly by printing-presses. The imitation stained glass now so frequently seen is the product of the old-fashioned hand printing-presses. Blocks of wood are used to convey the impression desired ; these blocks are inked with oil colours specially mixed for the purpose. Sheets of thin, porous, hand-made paper are used, prolonged impressions being given, so that the oil colours will thoroughly permeate the paper ; a separate impression is made for each colour. The design desired having been printed on as many sheets as are required to complete the pattern, which may be as large as a cathedral window if required, the sheets are soaked in warm water for half an hour, sponged off on being taken out, and coated on one side with a thin cement. A similar coat of cement is applied to the glass on which the printed paper is to be finally placed, the paper is laid over this, and the back varnished over. The glass thus becomes, to all appearances, stained glass. The effects of lead lines of irregular curves and fragments of coloured glass in mosaic are reproduced with as great brilliancy and as fine artistic effect as in genuine glass. Time and changes of temperature exercise no ill effects on the printed stained glass, which is frequently mistaken for the real and far more costly and just as destructible article.

Window Transparencies.—The process of preparing a transparent plate or panel for the reception of an illuminated design or ornamental sign consists in first coating the plate with a film of bichromatized gelatine, then subjecting it to the action of light, under a transparent plate or negative having the design or subject opaque and the ground transparent; then removing the bichromate alone, by means of water, and finally rolling up the gelatinized side of the plate with an opaque fatty ink or colour, which is allowed to set or dry, thus making the ground and shade-lines on the plate opaque, while leaving the design or subject transparent; and finally after the ink or colour is set or dry, backing or covering the plate or panel over the transparent or design portion with metal foil, pigment, or illuminating colours to make an ornamental plate, panel, or sign, or a transparency

Inside Coating for Paper Boxes.—A composition for coating the insides of paper boxes consists of the following ingredients :—Paraffin and bicarbonate of soda or alum, the proportion of the soda or alum being sufficient to harden the paraffin when applied to the box. These ingredients are to be thoroughly mingled by heating. The alum is not absolutely necessary, since the soda will harden the paraffin sufficiently to withstand the action of most liquids, paints, etc., but where considerable heat is to be withstood alum is used. There are several substances—such as glue and gum arabic —which may be added to the composition to render the latter more impervious to turpentine ; but this is not necessary, as the paraffin and soda are unaffected by turpentine or any of the usual solvents. Of course the soda combined with the paraffin has additional advantages beyond the mere hardening of the paraffin, since the oil and odour of the latter are entirely neutralized, and the box is thereby rendered suitable to contain liquids to be used as food.

White Marks on Blue Paper can be obtained by using a solution made by simply dissolving soda in water nearly to saturation, using it the same as ink, with either a steel pen for writing or a drawing pen for lines. The lines come out white instantly.

Imitation of Catgut.—Twine may be prepared so as to possess the appearance of catgut by treating it as follows : Soak the twine for one half hour in a solution of glue ; dry and soak for an hour or two in a strong infusion of oak bark, to which a little catechu has been added. Then dry the twine and smooth with a rag saturated with oil. The oak bark infusion may probably be replaced with advantage by a solution of bichromate of potassium.

Cheap Jacketing for Steam Pipes.—A recipe is as follows :—Wrap the pipe in asbestos paper, and lay a number of strips of wood lengthwise, from six to twelve, according to the size of the pipe, and bind them into position with the wire ; around the framework thus constructed wrap roofing paper, fastening it with paste or twine. If exposed to the weather, use tar paper or paint the outside.

To Re-Ink a Type-Writer Ribbon.—Some time ago (says a correspondent of the "Scientific American") I tried the experiment of re-inking a ribbon with such success that I never expect to buy one again. In two ounces or more of any ordinary writing fluid put a spoonful of thick gum-arabic mucilage and a teaspoonful of brown sugar, warm the mixture and immerse the ribbon long enough to become well saturated. When dry, spread the ribbon on a board and brush it well with glycerine. Should there be too much " colour " in the ribbon press it out, between papers, with a warm flat iron; or, if too dry, brush it again with glycerine. The secret of the ribbon giving out its colour is in the glycerine, and if you have body enough in the colour there is no danger that it cannot be made to work well. Such a ribbon is not affected by the dryness or humidity of the atmosphere, and I esteem mine much better than any obtained from the trade. It may be that I was fortunate in hitting upon just the right proportion of the different constituents, and possibly a second trial might not be so successful; but I think, with a little care, anyone could do as well with the same or similar means. My object was to get body to the " colour," hence the mucilage and sugar. Then it was necessary that the ribbon should retain a certain degree of moisture, for the gum and sugar make it dry and harsh, so the glycerine coat was put on; but there was danger of smearing the paper with too much moisture, or a wrinkled surface, and the ironing obviated this.

To Clean Brass Rules.—When verdigris gathers on the face of brass rule, and it won't print sharp, take a little diluted oxalic acid and wash the face—never scrape it with a knife.

Aluminium Silver.—This is made by melting together one part of silver with three or four of aluminium. It is valuable for articles in which one of the main objects is to obtain lightness, such as the instruments used for marine observations. Those parts of such instruments as octants and sextants which, if made with other metal, would weigh several pounds, will, when made of aluminium silver, only weigh one pound. The alloy can be turned and filed easily, which is not the case with the pure aluminium ; the latter is too soft, and has the objectionable property of sticking to the file.

Gloss Varnishing.—Some valuable information on this
subject was recently given. The question is of great importance to the trade. If a gloss will not stay on paper, and
it is known that the varnish has given good results before, of
course it can only be the fault of the paper; still there are
so many different kinds of gloss varnishes in the market, and
so many methods of varnishing, that we are sure the following
will be of interest to all our readers:—It is evident that all
lithographers who are not in the possession of a varnishing
machine would prefer to print a gloss upon the printed sheets.
Although as a general rule a printed gloss never can be as
good as a coated one, yet we shall speak of the best gloss
obtainable by printing and known only to a very few; but
before giving this receipt we have to go a little deeper into
the matter in order to make it clear why some varnishes are
better than others, and why some papers lose the gloss after
varnishing and printing: (1) the varnish can only stay on a
printed sheet if the sheet does not absorb the varnish; (2)
the paper absorbs the varnish when the sizing does not
contain glue enough. A regular writing or so-called sized
book paper is not always fit to print work on that has to
be varnished afterwards; the paper may be tested before
printing with the varnish to be used, but the fibre of the
paper has nothing to do with it; paper of any fibre, if sized
enough, will keep the gloss on its surface. It is the same
with enamelled papers. Only the coating of the paper is the
cause if the printed sheets keep the gloss or if the gloss
soaks into the paper. Every enamelled paper which lifts the
ink nicely in printing is the worst for varnishing. In short,
what is wanted is a paper into which the varnish does not
soak. Now of course if we have not the paper as desired, we
have to find a remedy, and this is very easily done. Take
for instance photographic views. Be sure to have the lightest
tint printed all over the picture, and print this tint with as
little thin varnish in it as possible; add as much silicate basic
potassa, drop by drop, until the tint becomes of the consistency of body colour, without using heavy white and
magnesia. This single tint will give to the paper a sizing,
and enable it to keep the gloss. Bear in mind that the more

colours are printed on the top of each other the better the gloss will be, and that when these facts are observed any gloss varnish in the market will remain on the paper. The best gloss varnish for printing, however, is forty parts of varnish No. 3, forty parts of damar varnish, fifteen parts of yellow beeswax, and about five parts of silicate basic potassa. Melt the wax first over a fire, then add the varnish (always stirring), and when everything is well mixed, which can be proved if, by putting a wooden stick into it, the varnish runs evenly and of uniform thickness, let it become perfectly cold, and do not add the water gloss until the varnish is to be used in printing. The gloss varnish, instead of being done by an extra printing, can also be mixed with any tint and printed at the same time ; it will remain on the paper, and will neither break nor crack.

To Paint with Oil and Water Colours upon Fine Art Terra Cotta.

—1. Sketch the design in pencil on cardboard or canvas. 2. Mix the oil colours with a body like flake white, and use some oil megilp in place of much turpentine. Lay this on fairly thick and let it dry for some time. 3. Then tint and finish with the colours, without the flake white, but still using megilp. 4. When quite dry put on a slight coat of spirit copal varnish. Pencil marks can be rubbed out with india-rubber. Spots of paint or mistakes can be removed by spirits of turpentine. *For Water Colours :* 1. Sketch the design in pencil and coat it with prepared size ; allow it to dry·perfectly. 2. Mix the colours with Chinese white, and use some wax water megilp in place of much water. 3. Then tint and finish with colours without Chinese white, still using wax water megilp. 4. When dry (if needed) put on a thin coat of spirit copal varnish. Black terra cotta does not require size.

Colouring Transparencies and Photographic Cards.

—The first thing to be done is to prepare the paints. Get a small quantity of different aniline colours or dyes, and dissolve them separately in spirits of wine, gradually adding the spirit until all is dissolved ; dilute by about its own bulk of water, and add ox-gall until the colours flow smoothly from a camel-hair brush over glazed paper; when this has been attained the colours are ready for use, and the painting may be commenced. It is advisable for the beginner to commence with a portion of the transparency which has the smallest surface of the same colour, as it requires a little practice to lay on an even coat on a large surface, such as the sky or sea. If the colour is piled on by degrees with dilute colour, it renders the laying on of a smoother coat much easier. It is as well to give the transparency a coat of varnish when the colouring is completed and quite dry. For colouring cards the colour should be laid on very dilute (especially the flesh colour), and in successive washes until the desired depth has been attained. The colours most useful are : lemon-yellow, green, orange, red, blue, and violet. For flesh colour, lemon-yellow and red; for different shades of green, lemon-yellow and blue ; for lilac and purple, violet and red ; for jewellery and fair hair, orange. The commonest water-colour brushes will do.

Black on White in Photography.—The question how photographic paper is made that will give black lines on a white ground at one operation is answered as follows :—The paper is first coated with a solution of perchloride of iron and tartaric acid, dried and exposed in the usual way behind the tracing. The light reduces the perchloride of iron to the protochloride. The print is then immersed in a solution of gallic acid, which turns the coating of perchloride of iron not acted upon by the light black, but does not affect the portions reduced by the light, hence, as the light cannot go through the black lines of the tracing, the sensitized surface under them blackens under the gallic acid. Lastly, the print is washed and dried. Owing to the powerful action of the gallic acid it is difficult to obtain clear whites.

Rust Stains on Nickel-Plating may be removed by thoroughly greasing, and, after several days, rubbing with a cloth moistened with water of ammonia. Any visible spots may then be moistened with dilute hydrochloric acid and

immediately rubbed dry. Washing and the use of polish
powder complete the process.

To Silver Glass.—Triturate 8 parts of nitrate of silver
in a porcelain capsule with water of ammonia gradually
added, until the liquid, which has at first become muddy,
becomes clear, avoiding, however, an excess of ammonia.
To the solution add 2 parts of powdered sulphate of ammonium, and then 700 parts of distilled water. Transfer the
liquid to a bottle of dark or non-actinic glass, in which it
may be kept for a long time. In another bottle dissolve 24
parts of pure grape sugar in 700 parts of distilled water, in
which 6 parts of pure caustic potassa had previously been
dissolved. When the solutions are to be used, mix equal
volumes of them, pour the mixture into the glass vessel to be
silvered, and move it about so that the whole of the surface
to be coated may be wetted by the liquor. In ten or twelve
minutes the coating will be completed. The process must be
repeated several times, until the coating attains the proper
thickness. Then the vessel is carefully rinsed with rain or
distilled water, and dried by exposure to air. Finally, the
coating is varnished, for which purpose a solution of equal
parts by weight of damar and ether is recommended.

Preparing Panels for Painting.—An excellent
method of preparing paper tissues, wooden tablets and the
like, for the purpose of painting, has been successfully
employed in Germany, the advantages being that in the use
of crayon or chalk the colours adhere better to the surface;
in water-colour work the moisture is long retained, so that
the colour tones remain unchanged during the work, and in
oil painting the colours easily mix, greatly facilitating the
work of beginners. The process is thus described:—One or
two coats of size are first applied to remove porosity, then a
thin layer of a paste of white lead (200 gr.) and boiled oil
(50 gr.) is put on. After drying for half-an-hour, fine cotton
dust is sprinkled on the surface from a sieve, and by striking
the piece on the back the fine cotton fibres are caused to
rise, forming a velvet-like surface. The material is left to

dry for two or three days, and brushed with a woollen brush, so as to depress the cotton somewhat; then a mixture of white lead (8 parts), gold lac (1 part), spirits of turpentine (1 part), and starch (1 part) is applied. This is equally distributed by passage between rollers with a caoutchouc surface. The cotton particles are raised again with a fine brush, and after two or three days drying the material is put into a bath of equal parts of alcohol and water.

Reproducing Old Prints.—Moisten the print to be reproduced with a solution of soda and soap until the ink has been sufficiently softened. Place the sheet on a smooth slab of ivory, cover it with a sheet of white paper previously saturated in soapy water, and rub it with a muller until an impression of the print has been struck off. Between five and six copies may be taken in this manner, which will be sufficient in most cases. Another method consists in applying a thin layer of paste, with which the original is covered, after which an ordinary roller is passed over it backward and forward. The ink adheres only to the black portions. The layer of paste is then carefully removed with a sponge, the original sheet is placed on a lithographic stone, the whole submitted several times to pressure, and the sheet finally wetted, when it may be taken off. The same way is then adopted as in printing from a lithographic stone. In a third method acids and gum are applied. The sheet is wetted with diluted acid to remove the sizing, enable the liquid to enter the pulp of the paper, and prevent the paper from absorbing the colour. A roller is then passed over the sheet, as in the second method. To prevent the new colour from combining with the acid, a thin layer of gum arabic is applied to the original. The latter may then be transferred to a stone.

To Copy Printed Matter.—To do this on any absorbent paper, damp the surface with a weak solution of acetate of iron, and press in an ordinary copying press. Old writing may also be copied on unsized paper, by wetting with a weak solution of sulphate of iron mixed with a small solution of sugar syrup.

To Restore Faded Photographs.—It has been discovered that faded photographs can be restored to their original colour by immersion in a dilute solution of bichloride of mercury till the yellowness disappears. If the photo is mounted the operation can be performed by placing it in close contact for a sufficient time with blotting paper well saturated with the bichloride. The process does not restore lost detail, but simply removes the sickly yellow colour and makes the pictures bright and clear, possessing after the operation a much warmer tone than they did originally.

An Easy Way to Emboss.—Here is a field for ingenuity which will afford room for development. Very fine results can be obtained at little expense, and it will be a novelty, as very few printers have done such work. Take a piece of six-ply card with smooth, white surface, just the size of the card you wish to emboss, and sketch the shape you desire with a pencil, cutting out the design in one piece with a sharp knife ; then trim the edge of the inside piece, so that it will play freely through the outside piece. Paste the outside or female die *firmly* on the back of a wood letter large enough to hold it, and the inside or male die very *lightly* to the same letter ; then lock up the letter and put it on the press, remove your rollers, make a good hard tympan, and after thoroughly pasting the surface of the *inside* die, take an impression and hold the platten on the impression until the paste has time to dry. On opening the press the under die leaves the wood letter on which it was *lightly* held and adheres to the tympan, leaving the outside die attached to the letter on the bed of the press. Then set the gauges, feed in the cards in the usual way, and proceed to emboss.

Horse-Power.—Eight man-power is equivalent to one horse-power, and this last is arrived at by the power or force which a horse generally exerts. It is compounded of his weight and muscular strength, and decreases with his speed. It is generally reckoned, in mechanical calculations, equal to 33,000 lbs. raised one foot high per minute ; and if continued throughout the day of 8 hours, amounts to 150 lbs. conveyed a distance of 20 miles, at a speed of $2\frac{1}{2}$ miles per hour.

Demands of Public Libraries.—By Act of Parliament five copies of each work, and of each subsequent edition, must be despatched to the libraries of the British Museum, London; Bodleian, Oxford; University College, Cambridge; Trinity College, Dublin; and Advocates', Edinburgh. They are generally all sent through one official agent in London, who gives a receipt for them. If by chance these copies for the libraries are delayed or overlooked, a demand is soon made for them, and must be complied with. Privately subscribed books of a limited number, not sold in the ordinary way, or advertised, are exempt from these demands.

A New Style of Window Decoration.—It must have occurred to many that very beautiful window decoration might be produced, if we could transfer chromo-lithographs to glass. The difficulty lies in the fact that many of these are printed at the back with ordinary letter-press. Further, the paper transferred, that covers the colour, must be rendered nearly transparent. This is the way in which such a thing might be done:—Coat the paper thinly with a clear mucilage of gum arabic, spread it out evenly on the glass plate, and let it dry. The paper may then be pared down with the greatest facility by means of a glove-maker's knife, a piece of thin flexible steel, three inches wide by five inches in length. At one end a handle is usually affixed, the other end being ground to a very fine edge. It is used somewhat after the manner of a plane, the plate being pressed down nearly level with the paper, and the edge of the blade presented somewhat obliquely to the stroke so as to cut smoothly. To make the paper translucent saturate it with castor oil, and cover the back with a second glass plate.

To Prevent Scaling of Boilers.—The latest specific is the gum of the eucalyptus globulus, which, it is said, has the effect of thoroughly removing the scales which form on boilers and preventing rust and pitting. The effect of this preparation will, it is expected, extend the period of the usefulness of boilers 100 or 150 per cent., and at the same time insure a very considerable saving in fuel, as scale is a non-conductor of heat. The process of manufacture is patented, but we are not aware if the gum is yet in the market.

To Soften Sealing-Wax Blots.—These may be softened by means of alcohol and then scraped away with ossa sepia. Sealing wax of all colours easily dissolves in strong alcohol, and forms a most excellent varnish for small ornaments.

To Distinguish Iron from Steel.—A German paper gives the following test to distinguish steel from iron :—Pour on the object to be tested a drop of nitric acid of 1.2 sp. gr., let it act for one minute, then rinse with water. On iron the acid will cause a whitish-grey, on steel a black stain.

To Fasten Transparencies to Glass.—Transparent showbills, if the paper is properly prepared, should be easily fastened to glass slightly moistened, but if any difficulty is experienced very fine white glue, or preferably clean parchment clippings, boiled in distilled water in glass or enamel until dissolved, must be applied very evenly with a soft hair brush to the face of the bill ; then press it on the glass and in a few minutes the bill will be firmly fixed. Glass may be fixed to glass in this way and the cement will bear a good deal of dry heat.

To Write on Cotton Sheeting.—Cotton sheeting on canvas or linen may be written upon without the ink spreading by applying to the fabric a preparation (gum arabic and water) and allowing it to dry, then pressing the place with a moderately hot iron. If the fabric is glazed or starchy it is best to wash out the starch before applying the preparation.

New Ink Eraser.—The newest thing in this line is blotting paper saturated with a solution of oxalic acid and dried. It not only absorbs the ink of a blot, but will remove

the blot itself if the ink does not contain indigo or aniline colour. It might be dangerous in removing signatures from important papers, but the trace of the writing will remain, and can be made legible by adding ferro-cyanide of potassium or gallic acid.

A Fine Lustrous Polish for Fine Cabinet Work.—Half a pint of linseed oil, half a pint of old ale, the white of an egg, and one ounce spirits of salt (muriatic acid). Shake well before using. A little to be applied to the face of a soft linen pad, and lightly rubbed for a minute or two over the article to be restored, which should first be rubbed off with an old handkerchief. It will keep any length of time.

Jute in Linen.—The presence of this may be ascertained by washing and treating with dilute chlorine, when the jute will become of a reddish colour and the linen white. Cotton in linen can be destroyed with sulphuric acid, leaving the linen uninjured.

How Gelatining is Done.—First, it is requisite that the work should be done in a room free from dust. The principal arrangement required for gelatine varnishing comprises several plates of heavy glass fixed in wooden frames and marked each with a distinctive number. Shelves should be built up against the wall, covered with a layer of pasteboard, and perfectly level, so that the glass plates may rest in a level position. This is all-important. The gelatine is broken into fragments, tied up in a clean sheet of linen, and placed in cold water until sufficiently soaked. After this it is placed in a pot of water over an alcohol lamp, the boiling water dissolving the gelatine and all impurities being kept back in the cloth. The result is a readily flowing solution, to which an equal volume of alcohol is to be added, as without such an addition the mass would soon harden and set unevenly. The more convenient proportions are two parts of gelatine and five parts of water, to which add three parts of alcohol. The vessel should be kept covered to prevent the alcohol from volatilizing. A tin vessel, with graduated scale, should be used so as to get uniform quantities of the fluid. Before being covered with the gelatine the glass plate should be oiled slightly to prevent the mass from adhering. From the cup with the graduated scale the gelatine is poured in a lukewarm condition, about the consistency of molasses, upon the glass plate. The glass plate is moved to and fro until it is covered in all its parts, after which it is returned to its place upon the shelf. In about fifteen minutes the picture to be covered should be moistened uniformly upon the reverse, and then placed upon the gelatine mass, which by this time is somewhat solid. Here the picture is allowed to remain for two or three days, when with a dull knife the mass may be ripped round the margin and the picture pulled off with its gloss coating. A varnishing with collodion has the effect of making the gelatine coating waterproof and more flexible without detracting from the transparency. Frames and glass plates should be carefully cleaned of all particles of gelatine before being used again.

"Autotype" Plates.—The method used is as follows : A plate, preferably of glass, is carefully coated with a solution of gelatine containing bichromate of potash. It is then dried, and an ordinary photographic negative is placed in contact with it and exposed to the action of the light, which hardens all the parts corresponding with the transparent parts of the negative or the dark parts of the picture. After the proper exposure the plate is washed in cold water to remove all the sensitizing material, and it is then dried. The gelatine surface will be found to have changed, so that it will act precisely like a lithographic stone ; when moistened the parts that were protected from light by the opaque parts of the negative absorb water, while the other parts remain dry. A roller charged with fatty ink is rolled over the plate, the ink adhering to the dry parts and being rejected by the parts that have absorbed water. Paper is now placed on the inked surface and subjected to pressure, when the design will be transferred to the paper. Then the moistening, inking, and pressure are repeated until the required number of copies has been produced.

Celluloid, also called parkesine, xylonite, and cellulose, is a preparation of cellulose either from wood, cotton waste, paper, or other similar substance. Woody fibre is prepared for this purpose also in a similar manner to the one used for producing paper pulp. The prepared fibre is steeped in a mixture of nitric acid and rectified sulphuric acid, 1 of the former acid, of 1.42 specific gravity, to 4 of the latter acid. It will be seen from this that it is similar to gun-cotton. After it has been properly treated with this mixture, it is drained and pressed by hydraulic pressure to remove excess of acid. The original inventor has patented some improvements in its manufacture. He treats the woody or other fibre, either pulped or not, with a solution of nitrate of zinc, chloride of zinc, or other similar compound. He says the nitrate of zinc quickly dissolves the fibre. At a proper temperature more fibre is added to the solution until a proper consistency is obtained. The operation is performed by properly constructed machinery, and the excess of zinc nitrate is removed by washing in water, alcohol, or other suitable vehicle. He forms the articles which are to be made from it whilst it is in the soluble state and the solvent is in it, removing the solvent by subsequent washing. Celluloid has been used for making blocks for printing from, and slips of it pasted on wooden mounts make excellent surfaces for printing ground tints from. To render it non-combustible, phosphates of lime, zinc, or alumina, oxalates of lime, magnesia, or zinc, baryta, soda, strontia, etc., are added in quantities amounting to from 10 per cent. to 150 per cent. When coloured with dyes or pigments, it is used for making small-tooth combs, billiard balls, knife handles, and a thousand and one other purposes. But one of the most novel is the use of it in making fronts, cuffs, and collars—when they become dirty all that is required being a little soap and water with which they are wiped with a sponge.

To Write on Terra Cotta.—You can write on terra cotta as easily as upon paper by dipping the clay tablet in milk with a few drops of acid added, and then allowing it to dry.

Chamois Leather.—It is a common error that these skins are derived from the " Chamois goat." The skins really are made from the flesh side of sheep-skins, soaked in lime water, and in a solution of sulphuric acid. Crude fish oil is poured over them, and they are afterwards carefully washed in a solution of potash. To restore their flexibility, plenty of washing in strong suds made of olive oil and potash with a little water, and afterwards laying between brown paper wet with glycerine, is a good plan.

Gilt Sunk and Raised Letters in Stone or Marble. —These letters require much more care and preparation than for printing. First a coating of size is laid on and then several successive coats thickened with finely powdered whiting until a good face is obtained. Each coat, as it becomes dry, must be rubbed down with fine glass paper before applying the next. Then go over it evenly and carefully with gold size and apply the gold leaf, burnishing with an agate; several coats of leaf—all burnished—will be required to give a good effect. A gold colour stain, consisting of equal parts of zinc sulphate, ammonium chloride, and copper acetate (verdigris), all in fine powder, is sometimes used for the same purpose, and carefully applied has a good appearance, though of course not equal to the gold.

To Remove Grease from Paint.—This may be removed from paint on machinery with benzine or naphtha, but it must be used quickly and carefully to avoid bringing off the paint as well.

An Electric Spark in a dusky atmosphere causes dust to settle, and if the air be smoky, clears it. This is probably one reason why the air seems so clear after a thunderstorm, even if little rain has fallen.

Spots on Engravings, etc.—A few drops of ammonia in a cupful of warm rain water, carefully applied with a sponge, will remove these.

To Mount Chromos for Framing.—First soak for fifteen minutes in a shallow dish, or lay between two newspapers that have been thoroughly saturated with water ; then paste to the panel of the wood or canvas which has been prepared to receive them ; care must be taken that there are no lumps in the paste.

Imitation Ivory.—Make isinglass and brandy into a paste with powdered eggshell, very finely ground. Give it any desired colour. Oil the moulds into which the paste must be poured, which must then be slightly warmed. Leave it in the mould until dry, when its appearance resembles ivory.

Painting Zinc.—A difficulty is often experienced in causing oil colours to adhere to sheet zinc. The employment of a mordant, so to speak, of the following composition is recommended :—One quart of chloride of copper, one of nitrate of copper, and one of sal ammoniac, are to be dissolved in 64 parts of water, to which solution is to be added one part of commercial hydrochloric acid. The sheets of zinc are to be brushed over with this liquid, which gives them a deep black colour. In the course of from twelve to twenty-four hours they become dry, and to their now dirty grey surface a coat of any oil colour will firmly adhere. Some sheets of zinc prepared in this way, and afterwards painted, have been found to entirely withstand all the atmospheric changes of winter and summer.

Electric Printing.—The outlines of the invention for electric printing are as follows :—The cotton, linen, or paper is first saturated with a solution of aniline black in water. The wet paper or fabric is then placed on a disc of metal, under which is a non-conductor, such as glass or caoutchouc. Another disc, on which is the drawing, is placed on top of the paper. The current is then connected with the two metal discs, and with a little pressure a photograph in black is produced. There are other modes of working, such as writing with iron or steel pencils, which will act as electrodes. The solution of aniline shows up the lines better, and does not blot if tragacanth, gelatine, or starch be added.

Copyright.—The term of copyright in any book published *in the lifetime of the author* shall endure for the natural life of such author, and for the further term of seven years, commencing at the time of his death; provided that, if the said term of seven years shall expire before the end of forty-two years from the first publication of the book, the copyright shall endure for forty-two years. The copyright in any book published *after the death of its author* shall endure for the term of forty-two years from the first publication thereof, and shall be the property of the proprietor of the author's manuscript from which it shall be first published.

Chromograph Bed for Copying Purposes.—Take white gelatine, 1 lb.; glycerine, 1 lb.; glucose, 1 lb.; strong white glue, 1 lb.; water, 5 lb. The glue should be placed in water for a short time, and then laid out to soak a little and allow the surplus water to run off. When this has been done, melt all in a vessel partly submerged in another vessel containing boiling water, care being taken that the ingredients do not get burned while cooking. As soon as all are thoroughly melted, pour gently through a fine sieve or a piece of coarse linen into a shallow tray the size desired. When properly cooled and allowed to season a day or so, it is ready for use. To make a writing copying-fluid for the " chromograph " process, take two parts of aniline violet to 30 parts of alcohol. Add water to same as may be needed.

Protection of Eyes in Proof Reading.—A piece of fine green glass laid on the proof is a relief to the eyesight.

How to Trace Photographs.—Lay over the photograph a sheet of gelatine, which may be obtained at an artists' colourman. With a sharp steel point scratch the outlines required, which will show light on a dark ground. Remove the gelatine and finely scrape some red chalk over it. Rub the chalk powder into the scratches, and dust off the superfluous powder with a brush. This tracing may then be laid down upon the stone and pulled through the press, when a good red chalk tracing may be obtained. It may also be laid upon the stone and rubbed down with the end of a pencil cut flat or other convenient tool being passed over it with some pressure. If wanted non-reversed upon paper, it is first passed through the press upon paper, to get a reverse, which in turn may be used to obtain the non-reversed one, for painting upon or other like use. If gelatine cannot be obtained, the photograph can be freed from grease by cleaning with india-rubber. The outlines may then be drawn with an ink composed of Indian ink, to which is added a small quantity of ox-gall, sugar, and gum. A piece of damp paper is then laid upon it, and the whole passed through the press. As the result is a reverse, tracing paper need not be used, because it is in the proper position for tracing to stone. Should a non-reversed tracing be required, use tracing paper for obtaining the set-off. A photograph thus treated will scarcely be soiled if the remainder of the Indian ink be washed off.

Hints for Mounting Photographs.—Binders are often called upon to insert photographs in books as illustrations. It has been said that a photograph cannot be mounted so that it shall lie flat ; this "cockling" has hitherto been a constant worry to the binder. The following are a few hints that may assist :—Silver-print photographs should never be placed on white mounts, because the high lights of the print are never pure white, and a bright white margin round the print spoils its beauty by killing its delicate half-tones. Lay the print on paper of various tones, and select the tint that harmonizes with it best. If starch is used, make it fresh every day, and in this way : take a teaspoonful of best starch in a large cup, add just enough cold water to break it up, but do not put in more than is quite necessary ; pour boiling water on it, stirring the while, until it is quite transparent ; when cold it may be used. Or if paste is used, take a teaspoonful of corn-flour, beat this well up in a teacupful of water till it is quite smooth and there are no lumps ; place this in a porridge saucepan (that is, a double one), and let it boil, stirring it continually ; it will turn to a thin and transparent paste that will be easy to work with and very adhesive. Gelatine dissolved in water—say a half-ounce to a teacupful—is also very good. It must be of the best quality and ought to be used hot. For mounting, three methods are given : *First*, after trimming the print all round, moisten it slightly (the object being to have it limp, without stretching it), by placing it between sheets of damp paper over night, and it will be about right next morning. Also slightly damp the mount, paste your print very carefully all over (using no more paste than is just necessary), lay it carefully on the mount, cover it with a piece of clean paper and rub it down well, and then place it in the standing-press between pieces of blotting paper, and allow it to dry under pressure. It may be perhaps necessary to take it out of the press and change the blotting paper. If all this be done properly, the photograph and its mount will lie quite flat. *Secondly*, another way is to paste the back of the print all over and allow it to dry ; damp the mount, lay the print on the damp mount and pass them through the rolling machine, or place them in the standing-press. *Thirdly*, take a piece of lithographic stone or a thick piece of glass ; glue this all over with the gelatine, place the print quickly down on the glued stone or glass, rubbing it smartly all over, then pick it up and lay it down on the mount. All this must be rapidly done, and, if properly so, a photograph can be easily mounted, even on thin paper.

· **Artificial Slate.**—A composition for the production of artificial slate consists of the following ingredients :—Powdered animal-bone, about fifteen parts ; white lead, about one part ; and linseed oil. These ingredients are prepared, mixed, and applied in the following manner :—Animal-bones, preferably bones from sheep, are burned in an open chamber or vessel to allow the free admission of air, to render them as white as possible. They are then ground very fine, and screened through a sieve. Of this fine powder, fifteen parts are mixed with about one part of white lead, with sufficient linseed oil to make it of the consistency of paint to be applied with a brush. The material—such as paper, pasteboard, linen, or any other material suitable to form the body of the artificial slate—is then painted on one or both sides with this composition

Y

several times until the desired thickness is obtained on the material, allowing each coating to dry before the next is applied. After the desired thickness of composition has been applied to the material, it is placed for several days in a dry, warm room, without the application of artificial heat. When thoroughly dry it is polished by means of pumice-stone dipped in linseed oil to a smoothness capable of being written upon by a lead pencil, and can thus be used similar to a slate. the pencil marks being at any time easily removed by the application of a little moisture. By using more or less white lead, the colour of this artificial slate may be varied, and any desired colouring matter may be added.

To Remove Rust from Polished Surfaces.—To remove rust and not scratch the finest polished surface, take of cyanide of potassium 15 grammes; soap, 15 grammes; chalk blanc de Meudon, 30 grammes; water sufficient to make a thick paste. As the removal of rust means removal of material surrounding rust pittings, however fine, a paste for the purpose must be a grinding material.

Mourning Paper.—A French machine puts the black edge on mourning paper before it is cut up into sheets. A brush applies the colour between parallel steel bands which are as far apart as the margin is to be wide, and these are adjustable. The bands are cleansed automatically to get off all the colour that is deposited on them by the brush.

On Paper Ruling.—There are several machines before the trade which enable any printer to rule a small quantity of paper without the delay and trouble of sending it to a regular paper ruler. In using these machines much difficulty is experienced sometimes in operating on some kinds of paper, not through any defect in the apparatus, but owing to a want of knowledge of the art itself. For instance, certain sorts of sized paper do not take the blue ink properly, that is to say, in ruling technicality it not does "strike." The lines are composed, in fact, of a series of bead-like dots, instead of being solid and unbroken. The amateur ruler should know that

paper requires a certain quantity of gall mixed with the ink. Ox gall should be procured if possible. Supposing you have prepared your ink to the proper tint required, mix well with each quart one tablespoonful of gall. Increase the quantity of gall as you find the paper hard or greasy ; bearing in mind that the flannel over the pens must be kept perfectly clean by being washed every other day in a little warm water with soap. If the acid in the blue ink has not been properly destroyed all the gall you may use will not prevent the bead-like lines. Particular care must be taken, likewise, that all the pens have the same bearing ; if not, the one that presses too heavily will give a thick line, whereas the one that hardly touches the paper will show a dotted broken one. Assuming that each pen is of the same length, begin by regulating the bearing of them by means of the regulator attached to the carriage of the machine. This carriage is the part that grips the slide wherein the pens are placed. Let the nibs of the pens be of the same angle, which must be secured by drawing them over the sand paper several times. Do not work your pens too much on the slant, but tolerably upright. The parts that carry the carriage will regulate this. Paste blue is sold by the ink makers, and is used thus :—Dip a brush into the paste and rub three or four times round the basin. If the ink has been standing all night, it will then be necessary to turn it backwards and forwards in two basins.

How to Print White Enamel Dials with Names and Numbers.—A good many processes have been proposed, and some of them patented, for effecting this. One of the most successful is the following:—The design or inscription is engraved on a copper plate and coated with a preparation containing the coloured enamel to be used. A sheet of prepared paper is then placed on the copper plate and the whole subjected to the action of a press of which the surfaces are heated. The paper on which the design is now printed is stripped from the copper plate and made to adhere to the article to be ornamented. Soaking the paper in water allows it to be taken off, the design remaining, which is baked so as to fix the fusible pigment on the coat of enamel.

A Hint for Gas-Burners.—A single plate of perforated zinc, about a foot square, suspended over a gas-jet, is said to retain the noxious emanations from burning gas, which, it is well known, destroys the bindings of books, tarnishes gilding, and vitiates the atmosphere for breathing.

Gilt Lettering on Glass.—Embossed ornamentation or lettering on glass, as done by hydrofluoric acid, can be closely imitated by painting in the design with rather thick dammar varnish, water-glass, Canada balsam, thinned with turpentine if required, or a solution of gum arabic. Whitish and ground glass effects may be obtained by finely grinding a little sugar of lead or sulphate of baryta in the vehicle used. Colours may be used if desired. Any gilding is done so as to cover the design.

New Gum Bottle.—The latest thing is a glass ball with a hole in the middle through which the handle of the brush passes and moves freely up and down. The glass ball projects from the top and rests in the upper part of the neck of the bottle and by its use dust is effectually excluded. In using the brush the glass ball remains a fixture on the handle.

Scenting Paper.—Complaints are frequent that in scenting programmes, etc., professional perfumers are much too fond of using patchouli and other sickly abominations. Nothing beats the simple good old-fashioned rose perfume, and nothing is simpler to make: of rose-leaves take one pound, of sandal wood half-a-pound, and of rose oil a quarter of an ounce. Mix well together. The old-fashioned lavender mixture for scenting paper is made as follows: lavender flowers, one pound; gum benzoin, a quarter of a pound; lavender oil, quarter of an ounce. This is a cheaper scent than rose and by some people liked better.

To Keep Unmounted Photos Flat.—A Belgian formula to prevent the curling up of unmounted photographic prints prescribes as follows:—After the final washing, plunge the prints in a solution made of water, one part, alcohol, four parts, and glycerine, three parts.

Printing Subscribers' Names.—A Frenchman has suggested an alteration in the ordinary method of printing the addresses of subscribers to newspapers and other periodicals. Instead of printing them in batches of about 20 on single sheets, which require checking for each issue, he prints them on a strip of paper like a ribbon, and they are afterwards perforated between the addresses. This ribbon is rolled up on a bobbin and the whole of the labels occupy but a small space, while none can be accidentally lost. Each is afterwards torn off and pasted on the wrappers. The plan seems to be something like that adopted by some of the omnibus companies for giving out tickets.

Colours of Government " Blue Books."

The English official colour is Blue.

„	French	„	„ „	Yellow.
„	German	„	„ „	White.
„	Austrian	„	„ „	Red.
„	Italian	„	„ „	Green.

French Bronzing.—The wonderfully well-coloured articles of bronze now seen in the market, showing all tints and adhering thoroughly to the metal, are made by a French process which is no longer a secret. The method of manufacture lies in the use of solutions of sulphides of antimony and arsenic, which are applied to the bronze or brass articles. After thoroughly washing them with water, they are thoroughly dried, and then the sulphide solution is applied with a brush. The best way is to begin with a dilute solution of ammonia, giving one application. After drying the coating is brushed, and then the dilute solution of sulphide of arsenic in ammonia is used, which produces a yellow colour. The oftener this sulphide of arsenic is applied the browner will be the colour, and a deep tint may thus be finally obtained. By solutions of sulphide of antimony, either in ammonia or in sulphide of ammonia, the tint is reddish, and it is possible in this way to obtain either the most delicate red or the deepest dark red. If some of the parts of the article are rubbed more strongly a high metallic lustre is obtained.

Purple Metal.—The pretty purple-tinted metal which has lately come on to the market in the form of ornamental nicknacks, is an alloy made by melting in a clean crucible equal parts of copper and antimony.

Type-Writer Press Copies.—Better press copies of type-writer work can be obtained than of pen work, and just as rapidly. The following instructions will insure success: First place an oil sheet in the letter-book; on this lay smoothly a damp cloth, then the tissue leaf of the book, and on the last place the letter to be copied. If the letter be written on one side only, lay another oil sheet on the back of the letter, and proceed as before. If the letter is written on both sides lay it on the book as directed above, and on the back of the letter turn down the next page of the copybook; upon that spread a damp cloth, and upon the cloth lay another oil sheet. It should be firmly pressed a minute or two. How damp to make the cloth and how long to permit the book to remain in the press, experience will demonstrate. In taking out the letters lay dry blotters between the pages of the book, and also between the letters just copied. As many letters can be copied at one time as desired.

Rusting of Bright Steel Goods.—This is due to the precipitation of moisture from the air. It may be obviated by keeping the surrounding air dry. A saucer of powdered quicklime placed in an ordinary showcase will usually suffice to prevent rusting of cutlery exhibited therein.

Transparent Coating for Prints.—It may be worth noting that a transparent coating, combined with a permanent glossy appearance, may be imparted to prints by mounting them on wet cardboard and applying a decoction composed of three ounces of white glue, eight ounces of soft water, half the white of an egg, ten drops of glycerine, and three grains of French chalk heated until thoroughly dissolved. It is not implied that a valuable print is improved by this or any other similar process, but the receipt may be useful in many cases where pecuniary value is not a very important factor.

Boiler Incrustation.—One of the most serious annoyances which printers, who use steam extensively, have to encounter, is the incrustation which always goes on within the boiler, at a rate depending mainly upon the amount of lime with which the water is impregnated. In some districts, as most people are aware, this is so great that natural springs and wells will speedily convert sticks, or leaves, or any other objects that may drop into them, into what are popularly called petrefactions, the objects, whatever they may be, being apparently converted into stone. This petrefaction of the inside of a boiler not only seriously reduces the working power of an engine, but is a prolific source of explosions, and is at all times a source of anxiety and expense.

Keeping the Hands Soft.—Many printers—compositors especially—whose hands have become rough and tough will be glad of hints how they can keep them in good condition. A little ammonia or borax in the water used to wash the hands with, and that water just lukewarm, will keep the skin clean and soft. A little oatmeal mixed with the water will whiten the hands. Many people use glycerine on their hands when they go to bed, wearing gloves to keep the bedding clean ; but glycerine makes some skins harsh and red. These people should rub their hands with dry oatmeal and wear gloves in bed. The best preparation for the hands at night is white of egg, with a grain of alum dissolved in it. The roughest and hardest hands can be made soft and white in a month's time by doctoring them a little at bedtime ; all the tools you need are a nailbrush, a bottle of ammonia, a box of powdered borax, and a little fine white sand to rub the stains off, or a cut of lemon, which will do even better, for the acid of the lemon will clean anything. There is no reason why a man should have hard and rough hands if with a little care he can have them soft and smooth. The hands are likely to be more dexterous if the skin is always pliable.

Plus and Equal Marks.—The signs + and −, it is said, were first used by Christopher Rudolf about 1524. The sign = was first employed by Robert Recorde in 1557, because, said he, " noe 2 thynges can be more equalle."

Painting Monograms.—To begin the painting of a monogram on paper, the first requirement is to know what colour the gears are to be striped, as the colours used in striping almost invariably govern the colour or colours used in the ornament, except when coats of arms—the colours of which are arbitrary—accompany the monogram, in which case the monogram is to be painted in its colours. When painting in relief—that is, in one colour, with its tints and shades, it is a good plan to first lay the whole design in with a medium shade of the colour, which gives an opportunity to lay in the shades with the dark shades of that colour, and then the lights with tints of the same colour, made by the addition of white. Many first lay in the whole design in gold or silver, and then glaze one letter with carmine, another with blue, another with verdigris, and so on. Where there are three letters two may be coloured, say Indian red glazed with carmine, and the prominent letter with vermilion, " cut up " with dark red, and high-lighted with vermilion and white.

" Perennial " Ink-Eraser.—A patented fluid for removing ink from paper or parchment—in order to rectify a mistake or clean off a blot without any injury to printers' ink, or ruling upon any mill-ruled paper, and leaving the paper or parchment as clean and good to write upon as it was before the mistake or blot was made—consists of one ounce of solution of chloride of lime combined with two drops of acetic acid. The end of the penholder is dipped into the fluid, and applied to the writing without rubbing. When the ink has disappeared the fluid is taken up with a blotter. To remove stains from laces, etc., the stained part is dipped into the fluid and then rinsed in clean water.

What Constitutes Cost ?—To the items of paper, composition, proof-reading, presswork, etc., add for rent and expenses, and interest upon investment in type, presses, etc., upon each job you do, 20%. This will give you about the *dead cost.* Now, if you want a profit, add one-third to the total, and in some cases one-half.

Reminders in Estimating.—Printers should accustom themselves to run down this list so that there be nothing omitted:—

Composition	Stitching or sewing
Presswork	Folding
Cards or paper	Cutting
Rulework	Woodcuts
Tabular work	Ornaments
Marginal or foot notes	Cutting brass rule
Colour work	Coloured ink and bronzes
More than one working	Purchase of peculiar sorts
Pressing	Numbering
Ruling	Stereotyping
Eyeletting	Electrotyping
Making up furniture, etc.	Night work

In making up an estimate always give yourself the benefit of a doubt, for when your price has once gone out to the parties calling for it there is no appeal, but your customer has always the right of asking for a revised estimate.

Steam Pipe for Heating.—The "Master Steam Fitter" gives the following rule for finding the superficial feet of steam pipe required to heat any building with steam: One superficial foot of steam pipe to six superficial feet of glass in the windows, or one superficial foot of steam pipe for every hundred square feet of wall, roof, or ceiling, or one square foot of steam pipe to eighty cubic feet of space. One cubic foot of boiler is required for every 1,500 cubic feet of space to be warmed. One horse-power boiler is sufficient to 40,000 cubic feet of space. Five cubic feet of steam, at seventy-five pounds pressure to the square inch, weighs one pound avoirdupois.

Preventing Explosion of Boilers.—A device has been patented by a German. It consists of a plate or cover held on a packing surrounding the outlet steam pipe, a weighted rod or stem holding the cover or plate on the packing, while a stop prevents the plate or cover seating itself on the outlet pipe after the packing is removed or thrown out by the pressure of steam from the boiler.

Power of Belting.—A belt travelling 800 feet per minute will safely transmit one horse-power for each inch in width if the pulleys are both of the same diameter, and the belt laps over one half of each ; but if the belt laps on but one quarter of either pulley's circumference, then it would have to travel 1,230 feet per minute to transmit a horse-power for each inch in width.

The Ampersand.—The "character and" or "short and" as it is known among printers, has the above title in the dictionaries, where it is said to be "a corruption of *and*, per se *and*, i.e. *and*, by itself *and*." It was originally formed—as may be seen in some old-style italic founts of to-day—of a combination of the capitals E and T, making the French and Latin word "et," signifying "and." Its only use is in connecting firm and corporation names, in &c., and it is sometimes permitted in display lines where the whole word cannot be inserted. Its use in any other case is obviously improper. Yet Smith, in his day, mourned that the contraction *&* was obliged to yield and to suffer "its comely form" to be supplied by the single letters e and t. In a French history of printing, published in 1740, the character *&* is used for *et* in every instance except where beginning a sentence. In giving the order of the lower-case sorts, Smith places the & between e and f.

The "Printers' Devil."—This trade term originated in Italy. Aldus Manutius was a printer in Venice. He owned a negro boy, who helped him in his office ; and some of his customers were superstitious enough to believe that the boy was an emissary of Satan. He was known all over the city as "the little black devil," from his dirty appearance, as his face and hands were generally well smudged with printing ink. Desiring to satisfy the curiosity of his patrons, Manutius one day exhibited the boy in the streets, and proclaimed as follows : "I, Aldus Manutius, Printer to the Holy Church and the Doge, have this day made public exposure of the Printers' Devil. All who think he is not flesh and blood may come and prick him!"

The Printers' Wayz-goose.—The origin of the word "wayzgoose" is not generally known. On the authority of Bailey the signification of the term is a "stubble-goose." Moxon, writing in 1683, gives an early example of its use in connection with the annual dinners of the printers of that time. He says:—"It is also customary for all the Journeymen to make every Year new Paper Windows, whether the old ones will serve again or no; Because, that day, they make them the Master Printer give them a *Way-goose;* that is, he makes them a good Feast, and not only entertains them at his own House, but, besides, gives them money to spend at the Alehouse or Tavern at Night; and to this Feast they invite the *Corrector, Founder, Smith, Joyner,* and *Inkmaker,* who all of them severally (except the *Corrector* in his own Civility) open their Purse-strings and add their Benevolence (which Workmen account their duty, because they generally chuse these Workmen) to the Master Printer: But from the *Corrector* they expect nothing, because, the Master Printer chusing him, the Workmen can do him no kindness. These Way-goose are always kept about Bartholomew-tide. And till the Master Printer hath given this *Way-goose* the journeymen do not chuse to work by Candle-Light." Other authors have quoted Moxon on the above, adding, however, riders of their own composition, more fully explaining the meaning of the term. Thus Timperley, writing in 1839, in a footnote, says:—"The derivation of this term is not generally known. It is from an old English word *Wayz,* stubble. A stubble-goose is a known dainty in our days. A wayz-goose was the head dish at the annual feasts of the forefathers of our fraternity." From this it would appear that the original derivation was from the goose which occupied the place of honour at the dinner, and not, as some have striven to show, from the excursion which usually forms part of their festival.

The Use of the Half-title.—The use of this fly-title, sometimes called a bastard title, is for the purpose of protecting the general or full title from injury. Without this additional leaf in front, the title-page, being the first in the book, would be very likely to get soiled.

Ancient Use of Vermilion.—Native cinnabar, or vermilion, a sulphuret of mercury, was, it is said, first prepared by Kallias, the Athenian, five hundred years before the Christian era. There was a minium or cinnabar wrought in Spain from stone mixed with silver sand; also in Colchis, where they disengaged it from the fronts of the high cliffs by shooting arrows at them. Pliny and Vitruvius call it minium, and Discorides observes that it was falsely thought by some to be the same as minium. Vermilion is the colour with which the statues of the gods were painted. It was abundant in Caramania, also in Ethiopia, and was held in honour among the Romans. Their heroes rode in triumph with their bodies painted with vermilion, and the faces of the statues of Jupiter were coloured with this pigment on festal days. The monochrome pictures of the ancients were wrought with it. There was also an artificial kind of cinnabar, a shining scarlet sand, from above Ephesus. Vitruvius and Pliny say that vermilion was injured by the light of the sun and moon. To prevent this result the colour was varnished by a mixture of wax and oil. Sir Humphry Davy found vermilion in the Baths of Titus.

Cultivating a Specialty.—All good work, if it is the very best of its kind, has come of late years from men who have made specialities of a particular kind of effort. Of printing this is as true as of any other art, and I find that but seldom does an office succeed in doing all kinds of work well. It is my theory that one office, or one head man of an office, should work exclusively on hand-presses, another on job-presses, a third on registered colour work, and another on high-class black and white, or single-colour work. Of course this is impracticable just now; but it is as certain to come in time as a distinguishing characteristic of all great offices as it is true that the indications unquestionably point that way even now.

Lead Pencils.—A perfect one should be strong, smooth, black, soft; should keep its point well in wear, be pleasant to use, and the mark should disappear entirely under rubber, leaving the paper perfectly clean.

Coats of Arms.—It will be in the recollection of many that the Government recently prohibited the use of the Royal Arms by persons not having written authority to display them. The same rule applies to the arms of the various counties and boroughs. Except by Royal licence no person has the right to assume any heraldic device, unless by sanction of Heralds' College and the payment of a yearly duty ; and no tradesman has any more right to use the coat of arms of the borough or county in which he resides, on anything he issues in the way of trade, than he has to print thereon the arms of the lord lieutenant of the county, which, of course, he would never think of doing. As this rule is frequently infringed, we give this information as a caution, else some persons may find themselves possessed of a large quantity of printed matter, against the issue of which they run the risk of receiving an injunction in Chancery, or some other equally disagreeable process at law.

Good Light.—There are few mechanical occupations where the need of good light is so imperative as in a printing office. There are so many small details that have to be seen to, whether it be in composition or presswork. Defective light of any kind greatly impairs the efficiency of every man compelled to labour in it. In this respect, it is very poor economy to locate a printing office in a room or rooms where the amplest light cannot be obtained, or to use a poor light. The best of oils and gas are poor substitutes for daylight, and besides, they cost more money. Too vivid a light can be shut off by screens or curtains ; but none of us can evoke the sunshine at will, or lengthen the day at either end. The continual burning of gas all day, as observed in some printing offices, is wasteful indeed, and the extra expense it entails, directly or indirectly, would go a good way towards paying a higher rent for better lighted quarters.

Preventing Gummed Stamps, etc., Sticking.—To keep postage stamps and other gummed articles from sticking together, rub over the head ; the natural oil on the hair oils them.

Gold Leaf Printing on Silk.—Take a fine brush of camel's hair and coat the silk or satin ribbon with a thin layer of silicate of potassa (water glass). Let it dry and then print in the usual way with gold ink. Grind burnt sienna as thickly as possible with No. 3 varnish and add the following reducers : 10 parts of yellow wax, 10 parts of Venetian turpentine, 25 parts of No. 2 varnish, and 5 parts of burnt linseed oil. Use this mixture when thoroughly blended and melted together as a regular reducer to the sienna ink ; print as usual, taking care that too much ink never gets upon the roller, and when the impression is made put on the gold leaf, Dutch metal, silver leaf, etc. Place a sheet of the finest glazed paper (such as the gold bronze is usually packed in) with the glossy side toward the gold and pull once again through the press, by which you can make another new impression as usual in gold leaf printing on paper. Let the impressions remain over for a day, then rub off the superfluous gold with soft, clean cotton dipped in soapstone, and the impression will appear clean, sharp, and solid.

Relief and Plain Stamping.—The following directions issued by Messrs. Baddeley Bros., of Chapel Works, Moor-lane, London, will be found useful :—

DIRECTIONS FOR PLAIN STAMPING.—Fasten the die in die-dish with gutta percha if it has no shank, if it has a shank the screw in die-dish will be sufficient to hold the die firmly ; cut two or three pieces of gummed cardboard, take one piece and damp the gum, placing it on the die with the gummed side upwards, then bring down the lever with a smart blow (the " dab " or piece of jagged steel being at the top, to which the gummed card will then adhere), then take the other pieces and treat in a similar manner. After striking them two or three times, take the dab from the press, with the card counterpart adhering, and then take a small piece of gutta percha and heat it, but not so as to burn, and place it on the cardboard, spreading it with the finger over the work—first damping the finger to prevent the gutta percha sticking to it ; replace the dab in press and place a thin piece of paper on the die and bring the lever down with a moderate pressure—a counterpart will be the result—then take a penknife and cut away the superfluous gutta percha close to the work. The guide board being fixed, insert pins therein to suit the position required of impression on paper or envelope, and bring down lever to give the blow.

DIRECTIONS FOR RELIEF STAMPING.—When the counterpart is made, mix the colour (on a slab, using the best pale paper varnish) to the substance of thin gum ; then take the die out of press and brush the colour well into it—exchange into right hand, face downwards, and wipe off the colour on to printing paper (about a quire of rather rough 4to) on to the bench. Let this be done regularly and systematically, so that paper is not wasted. The colour by this process is taken off the sur-face, but left in the engraved work. Place the die again in the press, insert paper, bring down the lever with a sharp blow, and the result will be a perfect impression. If the colour on slab and the brush should get dry, a few drops of turps will moisten it. To keep brushes moist wrap them in paper after using, to keep the air from them. To make the impression glisten or shine, mix the colour a day or two before using, and use no turps. Never let the colour remain on slab, as it will dry up. Keep in small gallipots covered, after being mixed up on slab.

DIRECTIONS FOR CAMEO WORK.—The same instructions to form counterpart as above. Relief colours will not do for cameo ; they must be bought ready mixed. Take a small portion on to the slab, distribute the roller well, and roll the face of die thoroughly; after which, place the paper in position, and strike as for plain and relief. The treatment of cameo stamp-ing is almost identical with letterpress printing. The rollers are the same material, and require the same attention ; and especial care must be taken that too much colour is not taken up on the roller, and that it is well distributed, to avoid clogging up the letters.

Printing on Vellum.

Printing on Vellum.—It is said that no difficulty will be experienced in printing on vellum if it is first sponged over with soapy water.

Printing on Wood.—There is a French invention for printing on wood by means of hot type. A guide regulates the degree of heat. The inventor claims that the impression is as neat as if obtained by lithography. By using a special ink cold type may be employed. Only one man is required to work the machine, which will print 350 boxes or 400 flat pieces of wood an hour.

Daguerreotype.—The practical inventor of photography was Daguerre, a Frenchman. In 1839 he published a descriptive treatise in French—afterwards translated into English, German, etc.—entitled : " History and Practice of Photogenic Drawing on the true principles of the Daguerreotype, with the new method of Dioramic Painting. Secrets purchased by the French Government, and by their command published for the benefit of the Arts and Manufactures." A later edition contains an addition to the title-page : " This process consists in the spontaneous Reproduction of the Image of Nature received in the Camera Obscura, not with their colour, but with great nicety in the gradation of shades."

Maxims for Printers.

It is better to remain idle than to work at a loss.

Genius is as rare in printing as in any other art.

Legitimate competition is a sign of life and health.

Do your work carefully, striving for constant improvement.

Follow copy, provided it is good, and never copy anything bad.

You cannot be a successful printer if the imprint of care and study is not upon brain and hands.

Preserve all specimens of good work that come into your possession, and spend your leisure time in their study.

Unless an apprentice is possessed of an ambition and determination to excel, the chances are that he will always be but a poor workman.

Skill in business, a well-earned reputation for uniformly superior work, a good financial credit, promptness, honourable and liberal dealing, correct and steady personal and business habits, are absolutely necessary concomitants of success.

No matter how good a printer you are, you will learn something new every day ; and in every job you do for a customer, study how you can improve it next time. Never let a poor or carelessly executed job go out of your office, no matter even if, by mistake in " estimating," or for any other reason, you may lose money on this particular one.

Study the work of first-class printers. A skilful workman has expended time, thought, and labour in its production.

It is not the grace or beauty of a single line that produces the result sought. The specimen must be judged as a whole.

Never curve a line where it would look better straight.

Do not crowd a job to put in a flourish or ornament.

Elaborate borders can only be used effectively by first-class workmen.

A plain rule border, with a neat corner, is more effective than a display border on a small card.

Ornament has to be kept strictly within the stern chasteness of taste, and permits of no extravagance of detail.

Ornament should always be subservient to its proper use. Any superfluity or preponderance destroys the proper effect.

Better do a good, plain job in black ink and one style of type, than an outrageous combination of fantastic ornaments in the glowing hues of the rainbow.

The use of ornaments requires a cultivated taste. They were intended to " light up," not smother ; to give an " airy grace," not detract; to do away with " monotony," not make a dreary waste.

Colour Blindness.

Colour Blindness.—This is much more common among printers than is generally supposed, if one is permitted to judge from jobs sent out. The land, especially during the holiday season, is flooded with abominations of tint and taste, with miserable chromos and calendars that are a disgrace to the art. When will craftsmen learn to avoid the delusions and pitfalls of colour, and assert the strict taste embodied in black and white ? Zebra-striped and rainbow-illuminated monstrosities will ever be a plague to the inventor, and are worthy only of some demented members of the paste-brush brigade. Printing gains nothing and loses much from such

z

violations of established rules, which in many instances proclaim the perpetrator upon a long-suffering public to be afflicted with distorted vision and colour blindness. Of course, we do not refer to the product of the true artist, which is always pleasing, but to the efforts of those who invariably go beyond their depths.

The First Printed Book was printed in 1455 from movable types, and is called the 42-line or "Mazarin" Bible. It was executed at Mentz, and is generally attributed to Gutenberg, though of quite recent years thought to be due to Peter Schoeffer.

The First Dated Book.—The first book printed with a date, and the first example of printing in colours, is a psalter, printed in Latin, on vellum, Faust & Schoeffer, at Mentz, in 1457. The initial letter commencing each paragraph is printed in red. It is supposed to have been the first psalter printed. The British Museum contains a copy of this ancient publication.

Registration at Stationers' Hall.—This is not a compulsory matter, but should anyone desire to bring an action for infringement of title or copyright it must then be registered (if this has not already been done), the date of publication being the criterion of priority. A five-shilling fee is exacted by the Stationers' Company when the application is made and the title of the volume duly entered. Very frequently the fact of registration is expressed at the bottom of the title-page, but this is really not necessary if the date is there.

Weights, Measures, etc.

AVOIRDUPOIS WEIGHT.

16 Drams ... ... = 1 Ounce.
16 Ounces ... ... = 1 Pound.
14 Pounds ... ... = 1 Stone.
28 Pounds ... ... = 1 Quarter of cwt.
4 Qrs. or 112 lbs. ... ... = 1 Hundredweight.
20 Hundredweight ... ... = 1 Ton.

One pound Avoirdupois contains 14 ozs. 11 dwts. 16 grains Troy ; or 114 lbs. Avoirdupois are equal to 175 lbs. Troy.

APOTHECARIES' WEIGHT.

20 Grains ... ... = 1 Scruple.
3 Scruples ... ... = 1 Dram.
8 Drams ... ... = 1 Ounce.
12 Ounces ... ... = 1 Pound.

The pound and ounce Apothecaries' weight are the same as in Troy weight, but the smaller divisions are different.

APOTHECARIES' MEASURE.

60 Minims ... ... = 1 Dram.
8 Drams ... ... = 1 Ounce.
20 Ounces ... ... = 1 Pint.
8 Pints ... ... = 1 Gallon.

STANDARD TROY WEIGHT.

4 Grains	= 1 Carat.
6 Carats or 24 grains	= 1 Pennyweight.
20 Pennyweights	= 1 Ounce.
12 Ounces	= 1 Pound.

One pound Troy is equal to 13 ozs. $2\frac{1}{2}$ drs. Avoirdupois : the former contains 5,760 grains, and the latter 7,000.

The standard for gold coins is 22 carats of fine gold and 2 carats of copper; for silver, is 11 ozs. 2 dwts. silver and 18 dwts. alloy, from which 66 shillings are coined.

WINE AND SPIRIT MEASURE.

4 Gills	= 1 Pint.
2 Pints	= 1 Quart.
4 Quarts	= 1 Gallon.
2 Gallons	= 1 Flagon.
10 Gallons	= 1 Anker of Brandy.
18 Gallons	= 1 Runlet.
$31\frac{1}{2}$ Gallons	= 1 Half Hogshead.
42 Gallons	= 1 Tierce.
63 Gallons	= 1 Hogshead.
84 Gallons	= 1 Puncheon.
2 Hogsheads or 126 galls.	= 1 Pipe or Butt.
2 Pipes or 252 galls. ...	= 1 Tun.

The Imperial Gallon, which is the Standard Measure of Capacity, contains $277\frac{1}{4}$ inches (very nearly), or 10 lbs. Avoirdupois of distilled water, at a moderate temperature.

ALE AND BEER MEASURE.

2 Pints	= 1 Quart.
4 Quarts	= 1 Gallon.
9 Gallons	= 1 Firkin of Ale or Beer.
2 Firkins or 18 gallons ...	= 1 Kilderkin.
2 Kilderkins or 36 gallons	= 1 Barrel.
$1\frac{1}{2}$ Barrels or 54 gallons...	= 1 Hogshead.
2 Barrels	= 1 Puncheon.
3 Barrels or 2 hhds. ...	= 1 Butt.

LONG MEASURE.

12 Inches	=	1 Foot.
3 Feet	=	1 Yard.
5 Feet	=	1 Pace (geomet.)
2 Yards or 6 Feet	=	1 Fathom.
5½ Yards, or 16½ Feet	=	1 Rod, Pole, or Perch.
4 Rods, or 22 Yards	=	1 Chain.
40 Rods, or 10 Chains	=	1 Furlong.
8 Furlongs, or 1,760 Yards ...	=	1 Mile.
3 Miles	=	1 League.
60 Geographical, or 69½ English Statute Miles ... ...}	=	1 Degree.
360 Degrees	=	{Circumference of the Earth.

Horses are measured by the hand of 4 inches. The chain is divided into 100 links.

LAND OR SQUARE MEASURE.

144 Square inches	=	1 Square Foot.
9 Square Feet...	=	1 Square Yard.
30¼ Square Yards	=	1 Square Pole or Perch.
16 Poles	=	1 Chain.
40 Rods, poles, or perches ...	=	1 Rood.
4 Roods, or 10 chains, or 160 rods	=	1 Acre of Land.
640 Acres	=	1 Square Mile.

MOTION.

60 Seconds	=	1 Minute.
60 Minutes	=	1 Degree.
30 Degrees	=	1 Sign of the Zodiac.
90 Degrees	=	1 Quadrant.
4 Quadrants or 360 Degrees	=	1 Great Circle.
360 Degrees of Motion	=	24 Hours of Time.
15 Degrees	=	1 Hour.
1 Degree	=	4 Minutes.

SOLID OR CUBIC MEASURE.

1,728 Inches...   ...   . = 1 Solid Foot.
27 Feet   ...   . = 1 Yard or Load.
40 Feet of unhewn, or 50 feet of⎫
     hewn timber...   ...   .⎭ = 1 Ton or Load.
42 Feet   ...   . = 1 Ton Shipping.

A cubic foot of water weighs 1,000 ounces Avoirdupois.

DRY MEASURE.

2 Pints ...   . = 1 Quart.
2 Quarts   ...   . = 1 Pottle.
2 Pottles or 4 Quarts ...   ... = 1 Gallon.
2 Gallons   ...   . = 1 Peck.
4 Pecks, or 8 Gallons ...   ... = 1 Bushel.
2 Bushels   ...   . = 1 Strike.
4 Bushels   ...   . = 1 Coomb.
2 Coombs or 8 Bushels   ... = 1 Quarter.
4 Quarters   ...   . = 1 Chaldron.
5 Quarters or 10 Coombs   ... = 1 Wey or Load.
2 Weys ...   . = 1 Last.

All kinds of Grain, Salt, etc., for which this measure is used, are to be stricken with a straight roller. The standard bushel is $19\frac{1}{2}$ inches diameter and $8\frac{1}{4}$ inches deep, and contains $2,218\frac{1}{3}$ cubic inches.

WOOL WEIGHT.

28 Pounds...   ...   . = 1 Tod.
240 Pounds...   ...   . = 1 Pack.

Wool is sold at per lb. The price per lb. in pence represents the value per pack in pounds sterling.

CLOTH MEASURE.

$2\frac{1}{4}$ Inches ...   . = 1 Nail.
4 Nails   ...   . = 1 Quarter.
3 Quarters   ...   . = 1 Flemish ell.
4 Quarters   ...   . = 1 Yard.
5 Quarters   ...   ... = 1 English ell.
6 Quarters   ...   ... = 1 French ell.

Metrical System, with English Equivalents.

MEASURES OF LENGTH.

The Metre is the Unit ; it is about $39\frac{1}{3}$ inches in length.

Metric Terms.		Metres.		Yds.	Inches.
Centimètre	=	$\frac{1}{100}$	=	0	0·39
Décimètre	=	$\frac{1}{10}$	=	0·	3·93
METRE	=	1	=	1	3·37
Décamètre	=	10	=	10	33·70
Hectomètre... ...	=	100	=	109	13·08
Kilomètre	=	1000	=	1093	22·30

MEASURES OF WEIGHT (*Avoirdupois*).

The Gramme is the Unit ; it is rather more than $\frac{1}{2}$ an ounce.

Metric Terms.		Grammes.		Ozs.	Drams.
Centigramme ...	=	$\frac{1}{100}$	=	0	0·0056
Décigramme... ...	=	$\frac{1}{10}$	=	0	0·0564
GRAMME	=	1	=	0	0·564
Décagramme ...	=	10	=	0	5·643
Hectogramme ...	=	100	=	3	8·438
Kilogramme	=	1000	=	35	4·38

MEASURES OF CAPACITY.

The Litre is the Unit ; it is nearly $1\frac{3}{4}$ pint.

Metric Terms.		Litres.		Galls.	Pints.
Centilitre	=	$\frac{1}{100}$	=	0	0·0176
Décilitre	=	$\frac{1}{10}$	=	0	0·1760
LITRE	=	1	=	0	1·760
Décalitre	=	10	=	2	1·607
Hectolitre	=	100	=	22	0·077
Kilolitre	=	1000	=	220	0·77

MEASURES OF SURFACE OR SQUARE MEASURE.
The Are is the Unit.

Metric Terms.		Ares.		Sq. Metre.		Acre.	Sq. Yds.	Inches.
Centiare	=	$\frac{1}{100}$	=	1	=	0	1·1960	00·14
ARE	=	1	=	100	=	0	119·6033	10·79
Hectare	=	100	=	10000	=	2	2280·3326	22·79

Dates of the Quarter Days.

Lady Day	... ...	25th March.
Midsummer Day	... ...	24th June.
Michaelmas Day	... ...	29th September.
Christmas Day ...	...	25th December.

Stamps.

BILLS OF EXCHANGE payable on demand for any amount, 1*d.* Ditto of any other kind, also PROMISSORY NOTES, not exceeding—£5, 1*d.* ; £10, 2*d.* ; £25, 3*d.* ; £50, 6*d.* ; £75, 9*d.* ; £100, 1*s.* ; every £100, and fractional part of £100, of such amount, 1*s.*

LEASE, or AGREEMENT FOR LEASE, of lands, tenements, etc., for 35 years and under, at a yearly rent not exceeding £5, 6*d.* ; £10, 1*s.* ; £15, 1*s.* 6*d.* ; £20, 2*s.* ; £25, 2*s.* 6*d.* ; £50, 5*s.* ; £75, 7*s.* 6*d.* ; £100, 10*s.* ; exceeding £100, for £50, or fractional part of £50, 5*s.*

CONVEYANCE OR TRANSFER.—Of Bank of England Stock, 7*s.* 9*d.* ; East India Company Stock, £1 10*s.* Of any debenture Stock or funded debt of any company or corporation, and colonial generally—for every £100, or fractional part of £100, 2*s.* 6*d.* Copy or Extract, the same duty as original, but not to exceed 1*s.*

AFFIDAVIT or statutory declaration, 2*s.* 6*d.*

AGREEMENT, or memorandum of agreement, under hand only, not otherwise charged, 6*d.*

APPOINTMENT OF A NEW TRUSTEE, and in execution of a power of property, not being by a will, 10*s.*

APPRAISEMENT or VALUATION of any estate or effects where amount of appraisement shall not exceed—£5, 3*d.* ; £10, 6*d.* ; £20, 1*s.* ; £30, 1*s.* 6*d.* ; £40, 2*s.* ; £50, 2*s.* 6*d.* ; £100, 5*s.* ; £200, 10*s.* ; £500, 15*s.* Exceeding £500, 20*s.*

Rates of Postage, Money Orders, Parcel Post, etc.

RATES OF POSTAGE.—Throughout the United Kingdom, for prepaid letters :—

Not exceeding 4 oz. ... ... 1*d.*

Exceeding 1 oz. but not exceeding 2 oz. 1½*d.*

   „   2   „   „   4   2*d.*

   „   4   „   „   6   2½*d.*

   „   6   „   „   8   3*d.*

   „   8   „   „ 10   3½*d.*

   „ 10   „   „ 12   4*d.*

   „ 12   „   „ 14   4½*d.*

And so on at the rate of ½*d.* for every additional 2 oz.

A letter posted unpaid is chargeable on delivery with double postage, and a letter posted insufficiently paid is charged double the deficiency.

For rates of Foreign Postage, see the Post Office Guide, which is published quarterly.

POST CARDS.—Post Cards available for transmission in the United Kingdom only, are sold at 5½*d.*, or of finer quality at 6*d.* per packet of 10. They can also be had in smaller numbers, or singly. Foreign post cards, 1*d.*, 1½*d.*, and 2*d.* each. Stout Reply Post Cards are sold at 1¼*d.* each. Thin Reply Post Cards are charged 1¼*d.* each.

The front side of Post Cards is for the address only.

INLAND BOOK POST.—The Book Post rate is one half-penny for every 2 oz., or fraction of 2 oz. Every book packet must be posted either without a cover, or in a cover entirely open at the ends.

POSTAGE ON INLAND REGISTERED NEWSPAPERS.—(*Prepaid Rate.*)—On each Registered Newspaper, whether posted singly or in a packet, the postage when prepaid is one half-penny ;

but a packet containing two or more Registered Newspapers is not chargeable with a higher postage than would be chargeable on a book packet of the same weight.

The postage must be prepaid either by an adhesive stamp or by the use of a stamped wrapper. Every newspaper, or packet of newspapers, must be posted either without a cover or in a cover entirely open at both ends.

No newspaper may contain an enclosure, or any writing of the nature of a letter.

Post Office Telegrams.—The charge for Telegrams throughout the United Kingdom is 6*d.* for 12 words, and ½*d.* for every additional word. Addresses are charged for. Postage stamps are used for payment, and must be affixed by the sender.

Money Orders for the United Kingdom.—Money Orders are granted in the United Kingdom at the following rates:—

For sums not exceeding ... ... £1, 2*d.*
For sums exceeding £1 and not exceeding £3, 3*d.*
„ „ „ £2 „ „ £4, 4*d.*
„ „ „ £4 „ „ £7, 5*d.*
„ „ „ £7 „ „ £10, 6*d.*

Postal Orders.—A new form of postal order, for fixed sums, is now issued; on those for 1*s.* and 1*s.* 6*d.* the charge is ½*d.*; for 2*s.*, 2*s.* 6*d.*, 3*s.*, 3*s.* 6*d.*, 4*s.*, 4*s.* 6*d.*, 5*s.*, 7*s.* 6*d.*, 10*s.*, and 10*s.* 6*d.*, it is 1*d.*; for 15*s.*, and 20*s.*, it is 1½*d.* These notes can pass from hand to hand like money.

Money Orders Payable Abroad.—Money Orders are issued in the United Kingdom on France, Belgium, Switzerland, Germany, the United States, and several other foreign countries, and on most of our Colonies, at the following rates:—For sums not exceeding £2, 6*d.*; £5, 1*s.*; £7, 1*s.* 6*d.*; £10, 2*s.*

Registration.—On the prepayment of a fee of twopence, any letter, newspaper, or book packet may be registered to any place in the United Kingdom or the British Colonies. Every letter, etc., to be registered must be given to an agent of the Post Office, and a receipt obtained for it. Registered letter envelopes are sold at all Post Offices.

PARCEL POST.—Parcels not exceeding 11 lb. in weight may now be transmitted by the Inland Parcels Post under the following general conditions :—

The rate of postage is for an Inland Postal Parcel—

Not exceeding 1 lb. ... ... 3*d.*
Exceeding 1 lb. and not exceeding 2 lbs. 4½*d.*

And so on, adding 1½*d.* for every additional lb. up to 11 lb., which is charged 1*s.* 6*d.*

The dimensions allowed for an Inland Postal Parcel will be :—

Greatest length ... ... 3 feet 6 inches.
Greatest length and girth combined 6 feet.

For example, a parcel measuring 3 ft. 6 in. in its longest dimension may measure as much as 2 ft. 6 in. in girth, *i.e.*, around its thickest part.

A Parcel Post is now established between the United Kingdom and many foreign countries, and the British Colonies and Possessions generally.

STAMPS (POSTAGE AND INLAND REVENUE).—Postage stamps of the value of 1*d.*, 2*d.*, 3*d.*, 6*d.*, 9*d.*, 1*s.*, and 2*s.* 6*d.* are now used for Inland Revenue purposes to denote the duties on Agreements, Bills of Exchange, Delivery Orders, Receipts, Voting Papers, etc.

Days of Grace.

—Bills of Exchange or Promissory Notes, payable at any time after date, have three Days of Grace allowed: thus, a bill dated 1st January at two months' date is not due till 4th March ; but by a recent Act no Days of Grace are allowed on Bills drawn at sight or on demand ; such must, therefore, be paid on presentation. Bills falling due upon " Bank Holidays " are payable the day after ; but those falling due on Sundays, on Good Friday, or Christmas Day, must be paid the Day before.

Coins of Foreign Countries.

With approximate value in English money.

	£	s.	d.
AMERICA—Dollar = 100 Cents ...	o	4	1
AUSTRIA—Florin = 100 Kreuzers	o	1	7
CHINA—Tael of Silver	o	4	6
Dollar	o	3	1
FRANCE AND BELGIUM—			
Napoleon = 20 Francs ...	o	15	10
Franc = 100 Centimes ...	o	o	$9\frac{1}{2}$
GERMAN EMPIRE—20 Marks ...	o	19	7
Mark = 100 Pfennige	o	o	$11\frac{3}{4}$
GREECE—20 Drachma	o	14	2
Drachma = 100 Leptas ...	o	o	$8\frac{1}{2}$
HOLLAND—Ducat	o	9	5
Gulden or Florin = 100 Cents	o	1	8
INDIA—Mohur = 15 Rupees ...	1	10	o
Rupee = 16 Annas...	o	1	5
Anna = 12 Pies	o	o	$1\frac{1}{2}$
ITALY—Lira = 100 Centesimi...	o	o	$9\frac{1}{2}$
JAPAN—Yen = 100 Sen...	o	3	6
NORWAY, SWEDEN, AND DENMARK—			
Crown = 100 Ore	o	1	$0\frac{1}{2}$
PORTUGAL—Milreis = 1,000 Reis	o	4	2
RUSSIA—Imperial	o	16	$3\frac{1}{2}$
Rouble = 100 Copecks	o	3	$1\frac{1}{2}$
SPAIN—5 Dollar	1	o	6
Dollar = 20 Reals	o	3	10
Peseta (100 Centavos)	o	o	$9\frac{1}{4}$
TURKEY—Medjidié = 100 Piastres...	o	18	o
Piastre = 40 Paras...	o	o	2

Interest Table.—The following is a simple method of calculating interest on any given sum for any number of days, at five per cent. :—

Multiply POUNDS by the number of days for which it is required to ascertain the interest ; this sum, divided by 365, will give the interest in shillings at five per cent.

EXAMPLE.—Wanted to ascertain the interest on £479 for 71 days :—479 × 71 ÷ 365 = 93·17 = £4 13s. 2d. Or the interest on £250 for 73 days at the same rate :—

$$
\begin{array}{r}
250 \\
73 \\
\hline
750 \\
1750 \\
\end{array}
$$

$$365 \overline{\smash{\big)}\,18250} \left(50s. = £2\ 10s.\right.$$
$$1825$$

If any other rate is required, it is easily calculated by adding to or deducting from the five per cent. product :—

2½ per cent. is one-half.

3 per cent. is six-tenths.

3½ per cent. is seven-tenths.

4 per cent. is four-fifths.

6 per cent. is six-fifths.

7½ per cent. is one-half more.

Thus 5 per cent. upon £60 for ten months would be £2 10s. ; 2½ per cent., £1 5s. ; at 3 per cent., £1 10s. ; at 3½ per cent., £1 15s. ; at 4 per cent., £2. If the rate of interest be more than 5 per cent., then the addition must be added. Thus, to reckon 6¼ per cent., add one-fourth ; for 7½ per cent., one-half. Bankers and money-dealers calculate the interest for every day, and have volumes of tables constructed specially for the purpose, the five per cent. tables alone extending to nearly 400 pages.

INDEX OF SUBJECTS.

The entries in small capitals are those of sections.

CHISWICK PRESS:—C. WHITTINGHAM AND CO., TOOKS COURT,
CHANCERY LANE.

THE

BRITISH PRINTER

THE ACKNOWLEDGED TECHNICAL & ARTISTIC
EDUCATOR OF THE CRAFT.

CONDUCTED BY ROBERT HILTON.

THE

LARGEST SUBSCRIBED CIRCULATION

AND THE

LARGEST ADVERTISING PATRONAGE

OF ANY PRINTING AND PAPER TRADES' JOURNAL
IN THE UNITED KINGDOM.

Bi-Monthly. **9000** *each issue. Five Shillings a year.*
Specimen copy, Tenpence post free.

LONDON:

RAITHBY, LAWRENCE AND CO., LTD.,
25, PILGRIM STREET, LUDGATE HILL, E.C.

TECHNOLOGICAL HANDBOOKS.

" The excellent series of technical handbooks."—*Textile Manufacturer.*
" The admirable series of technological handbooks."—*British Journal of Commerce.*
" Messrs. Bell's excellent technical series."—*Manchester Guardian.*

EDITED BY SIR H. TRUEMAN WOOD.

A Series of Technical Manuals for the use of Workmen and others
practically interested in the Industrial Arts, and specially adapted for
Candidates in the Examinations of the City Guilds Institute.
Illustrated and uniformly printed in small post 8vo.

DYEING AND TISSUE-PRINTING. By WILLIAM
CROOKES, F.R.S., V.P.C.S. 5*s*.

" Whether viewed in connection with the examination room or the dye-house, the
volume is one which deserves a work of welcome."—*Academy.*
" The only previous qualification of which the student is assumed to be possessed is
an elementary knowledge of chemistry such as may be acquired from almost any of the
rudimentary treatises on that science. The author, building upon this foundation, seeks
to explain the principles of the art from a practical rather than from a theoretical point of
view. From the very outset he endeavours to explain everything with which the learner
might be puzzled."—*Chemical News.*

GLASS MANUFACTURE. INTRODUCTORY ESSAY by
H. J. POWELL, B.A. (Whitefriars Glass Works); CROWN AND
SHEET GLASS, by HENRY CHANCE, M.A. (Chance Bros., Birming-
ham); PLATE GLASS, by H. G. HARRIS, Assoc. Memb. Inst. C.E.
3*s.* 6*d.*

*COTTON SPINNING: Its Development, Principles, and
Practice.* With an Appendix on Steam Engines and Boilers. By
R. MARSDEN, Editor of the "Textile Manufacturer," and Examiner for
the City and Guilds of London Institute. Fourth Edition. 6*s.* 6*d.*

CONTENTS.—Introductory—Cotton—The Mill—Manipulation of the Material—Card-
ing and Combing—Drawing, Stubbing, and Roving—Development of Spinning—The
Modern System of Spinning—The Modern Mule—Throstle and Ring Spinning ; Doubling
—Miscellanea—Appendix.

" An admirable work on the subject."—*Manchester Examiner and Times.*
" Practical spinners, of whom Mr. Marsden is evidently one, will value this volume as
a handbook, and learners will find the fullest information given with the greatest possible
clearness."—*Manchester Courier.*

COTTON WEAVING. By R. MARSDEN, Examiner to the
City and Guilds of London Institute, Author of "Cotton Spinning."
With numerous illustrations. [*In preparation.*

COAL-TAR COLOURS, The Chemistry of. With special
reference to their application to Dyeing, &c. By DR. R. BENE-
DIKT, Professor of Chemistry in the University of Vienna. Translated from the German by E. KNECHT, Ph.D., Head Master of the
Chemistry and Dyeing Department in the Technical College, Brad-
ford. 2nd Edition, Revised and Enlarged. 6*s.* 6*d.*

" The original work is popular in Germany, and the translation ought to be equally
appreciated here, not only by students of organic chemistry, but by all who are practically
concerned in the dyeing and printing of textile fabrics."—*The Athenæum.*
" The volume contains, in a little space, a vast amount of most useful information
classified in such a manner as to show clearly and distinctly the chief characteristics of
each colouring matter, and the relationship existing between one series of compounds and
another."—*Journal of the Society of Dyers and Colourists.*

WOOLLEN AND WORSTED CLOTH MANUFAC-
TURE. By Professor ROBERTS BEAUMONT, Textile Industries
Department of the Yorkshire College, Leeds. Second Edition,
Revised. 7s. 6d.

CONTENTS.—Materials—Woollen Thread Manufacture—Worsted Thread Construction
—Yarns and Fancy Twist Threads—Loom-Mounting, or Preparation of the Yarns for the
Loom—The Principles of Cloth Construction—Fundamental Weaves—Hand Looms—
Power Looms—Weave-Combinations—Drafting—Pattern Design—Colour applied to
Twilled and Fancy Weaves— Backed and Double Cloths—Analysis of Cloths and Calcula-
tions—Cloth Finishing.

"The book is a satisfactory and instructive addition to the Messieurs Bell's excellent
technical series."—*Manchester Guardian.*

"It should be studied and inwardly digested by every student of the textile arts."—
Textile Recorder.

"A valuable contribution to technological literature."—*Irish Textile Journal.*

"The latest addition to the admirable series of technological handbooks in course of
publication by Messrs. Bell and Sons is a most valuable work, and will take at once a very
high place among technical manuals."—*British Journal of Commerce.*

PRINTING. A Practical Treatise on the Art of Typography
as applied more particularly to the Printing of Books. By C. T. JACOBI,
Manager of the Chiswick Press: Examiner in Typography to the
City and Guilds of London Institute. With upwards of 150 Illustrations, many useful Tables, and Glossarial Index of Technical Terms
and Phrases. 5s.

"The work of a man who understands the subject on which he is writing, and is able
to express his meaning clearly. Mr. Jacobi may further be complimented on having
supplied an excellent index."—*Athenæum.*

"A practical treatise of more than common value. . . . This is a thorough, concise,
and intelligible book, written with obvious mastery of all details of the subject."—*The
Speaker.*

"Mr. Jacobi goes into the minutest particulars . . . contains a large amount of information which will prove interesting to anyone who has ever had occasion to look into
a printed book or newspaper."—*Saturday Review.*

"It deals with the subject in an exhaustive and succinct manner. . . . We wish it all
the success it deserves in its efforts on behalf of technological education."—*Printing
Times and Lithographer.*

"There is much about it which pleases us. . . . It is well printed and well illustrated.
. . . He has written tersely and to the point."—*Printers' Register.*

"'Printing' is a book that we can recommend to our readers. It is literally full of
items which will be of importance to the printer in his daily toil."—*Effective Advertiser.*

BOOKBINDING. A Practical Treatise on the Art. By J. W.
ZAEHNSDORF. With 8 coloured Plates and numerous diagrams.
Second Edition, Revised. 5s.

"No more competent writer upon his art could have been found. . . . An excellent
example of a technical text-book."—*Industries.*

"To professional as well as amateur binders it may confidently be recommended."—
Paper and Printing Trades Journal.

"Its phraseology is simple, straightforward and clear, its arrangement systematic,
and its completeness apparently without a flaw."—*Guardian.*

PLUMBING. Its Theory and Practice. By S. STEVENS
HELLYER. With numerous illustrations. 5s. [*Immediately.*

SILK-FINISHING. By G. H. HURST, F.C.S. [*In the press.*

"THE SPECIALISTS' SERIES."

A New Series of Handbooks for Students and Practical Engineers.
Crown 8vo, cloth. With many Illustrations.

ELECTRIC TRANSMISSION OF ENERGY, and its
Transformation, Subdivision, and Distribution. A Practical Handbook by GISBERT KAPP, C.E., Member of the Council of the Institution of Electrical Engineers, &c. With numerous Illustrations. Third Edition, thoroughly revised and enlarged. 7s. 6d.

"We have looked at this book more from the commercial than the scientific point of view, because the future of electrical transmission of energy depends upon the enterprise of commercial men and not so much upon men of science. The latter have carried their work to a point, as is admirably shown by Mr. Kapp in his work, where the former should take hold."—*Engineer*.
"The book is one of the most interesting and valuable that has appeared for some time."—*Saturday Review*.
"We cannot speak too highly of this admirable book, and we trust future editions will follow in rapid succession."—*Electrical Review*.

HYDRAULIC MOTORS: *Turbines and Pressure Engines.*
For the use of Engineers, Manufacturers, and Students. By G. R. BODMER, A.M. Inst. C.E. With numerous Illustrations. 14s.

"A distinct acquisition to our technical literature."—*Engineering*.
"The best text-book we have seen on a little-known subject."—*The Marine Engineer*.
"Mr. Bodmer's work forms a very complete and clear treatise on the subject of hydraulic motors other than ordinary water-wheels, and is fully up to date."—*Industries*.
"A contribution of standard value to the library of the hydraulic engineer."—*Athenæum*.

THE TELEPHONE. By W. H. PREECE, F.R.S., and J.
MAIER, Ph.D. With 290 Illustrations, Appendix, Tables, and full Index. 12s. 6d.

Mr. Rothen, Director of the Swiss Telegraphs, the greatest authority on Telephones on the Continent, writes :—"Your book is the most complete work on the subject which has as yet appeared ; it is, and will be for a long time to come, *the* book of reference for the profession."
"Messrs. Preece and Maier's book is the most comprehensive of the kind, and it is certain to take its place as the standard work on the subject."—*Electrical Review*.

ON THE CONVERSION OF HEAT INTO WORK.
A Practical Handbook on Heat-Engines. By WILLIAM ANDERSON, M. Inst. C.E. With 61 Illustrations. Second Edition, revised and enlarged. 6s.

"We have no hesitation in saying there are young engineers—and a good many old engineers too—who can read this book, not only with profit, but pleasure ; and this is more than can be said of most works on heat."—*The Engineer*.
"The volume bristles from beginning to end with practical examples culled from every department of technology. In these days of rapid book-making it is quite refreshing to read through a work like this, having originality of treatment stamped on every page."—*Electrical Review*.

ALTERNATING CURRENTS OF ELECTRICITY.
By THOMAS H. BLAKESLEY, M.A., M. Inst. C.E. Second Edition, enlarged. 4s. 6d.

"It is written with great clearness and compactness of statement, and well maintains the character of the series of books with which it is now associated."—*Electrician*.
"A valuable contribution to the literature of alternating currents."—*Electrical Engineer*.

BALLOONING: A Concise Sketch of its History and Principles. From the best sources, Continental and English. By G. MAY. With Illustrations. 2s. 6d.

"Mr. May gives a clear idea of all the experiments and improvements in aëro-navigation from its beginning, and the various useful purposes to which it has been applied."—*Contemporary Review.*

SEWAGE TREATMENT, PURIFICATION, AND UTILIZATION. A Practical Manual for the Use of Corporations, Local Boards, Medical Officers of Health, Inspectors of Nuisances, Chemists, Manufacturers, Riparian Owners, Engineers, and Ratepayers. By J. W. SLATER, F.E.S., Editor of "Journal of Science." With Illustrations. 6s.

"The writer in addition to a calm and dispassionate view of the situation, gives two chapters on 'Legislation' and 'Sewage Patents.'"—*Spectator.*

A TREATISE ON MANURES; or, the Philosophy of Manuring. A Practical Handbook for the Agriculturist, Manufacturer, and Student. By A. B. GRIFFITHS, Ph.D., F.R.S. (Edin.), F.C.S. 7s. 6d.

"We gladly welcome its appearance as supplying a want long felt in agricultural literature, and recommend every farmer and agricultural student to possess himself with a copy without delay."—*Farm and Home.*

COLOUR IN WOVEN DESIGN. By Professor ROBERTS BEAUMONT, of the Textile Industries Department, The Yorkshire College. With 32 Coloured Plates and numerous Illustrations. 21s.

"An excellent work on the application of colour to woven design."—*Textile Manufacturer.*
"The illustrations are the finest of the kind we have yet come across, and the publishers are to be congratulated on the general excellence of the work."—*Textile Mercury.*

Works in Preparation—

LIGHTNING CONDUCTORS AND LIGHTNING GUARDS. By Professor OLIVER J. LODGE, D.Sc., F.R.S., M.Inst.C.E. With numerous Illustrations. [*In the press.*

THE DYNAMO. By C. C. HAWKINS, A.M.I.C.E., and J. WALLIS. [*Preparing.*

CABLES AND CABLE LAYING. By STUART A. RUSSELL, A.M.Inst.C.E. [*Preparing.*

THE ALKALI-MAKERS' HANDBOOK. By Professor Dr. GEORGE LUNGE and Dr. FERDINAND HURTER. Second Edition, revised, and in great part rewritten. [*In the press.*

ARC AND GLOW LAMPS. New and Revised Edition. [*Preparing.*

THE DRAINAGE OF HABITABLE HOUSES. By W. LEE BEARDMORE, A.M.Inst.C.E., Hon. Sec. to the Civil and Mechanical Engineers' Society.

LONDON: GEORGE BELL & SONS, 4, YORK STREET, COVENT GARDEN,
AND WHITTAKER & CO., PATERNOSTER SQUARE.

www.ingramcontent.com/pod-product-compliance
Lightning Source LLC
Chambersburg PA
CBHW021359210326
41599CB00011B/935